Touring the Climate Crisis

Touring the Climate Crisis

Saving the Earth around the World

Osseily Hanna

ROWMAN & LITTLEFIELD
Lanham • Boulder • New York • London

Published by Rowman & Littlefield
An imprint of The Rowman & Littlefield Publishing Group, Inc.
4501 Forbes Boulevard, Suite 200, Lanham, Maryland 20706
www.rowman.com

6 Tinworth Street, London SE11 5AL, United Kingdom

British Library Cataloguing in Publication Information Available

Library of Congress Cataloging-in-Publication Data

Names: Hanna, Osseily, 1978– author.
Title: Touring the climate crisis : saving the earth around the world /
  Osseily Hanna.
Description: Lanham : Rowman & Littlefield, 2021. | Includes
  bibliographical references and index. | Summary: "Author Osseily Hanna
  documents the links between climate change, social / wealth
  inequalities, the Global North / Global South, and renewable / fossil
  energy. In total, he visited 32 countries during his 6 year journey
  around the world. As well as photographs taken by the author, this book
  includes the testimonies of people who are affected by or fighting
  climate change, and who have devised ingenious local solutions to it" —
  Provided by publisher.
Identifiers: LCCN 2020047793 (print) | LCCN 2020047794 (ebook) | ISBN
  9781538149461 (cloth) | ISBN 9781538149478 (ebook)
Subjects: LCSH: Climatic changes. | Climatic change—Social aspects. |
  Climatic changes—Economic aspects.
Classification: LCC QC903 .H3676 2021 (print) | LCC QC903 (ebook) | DDC
  363.738/74—dc23

LC record available at https://lccn.loc.gov/2020047793
LC ebook record available at https://lccn.loc.gov/2020047794

To Daniel Miranda Silva

Contents

Part IV: Asia

Part V: Journey's End

Acknowledgments

$\mathscr{I}$would like to thank Suzanne Staszak-Silva, executive editor at Rowman & Littlefield, for her support and guidance; Lidija Mavra, PhD, line editor, for delving so deeply into the manuscript and enabling me to see my ideas more clearly; Joanne Foster, copy editor, for her conscientious editing; Melissa McNitt, production editor, for managing the delivery time line so effectively; and Nicholas Lalvani, who gave feedback on an early draft of the introduction.

My deepest gratitude goes to all of the interviewees in the book: Thomas Ayshford, Moeketsi Baloe, Vanesa de Blas, Felix Camposeco, Maria Ceron, Favio Chávez, Carlos, Chloe, Paty Fuentes, Jaime, Associate Professor Martin Hansen, Soh Horie, Lars Jacobsen, Ibrahim Kioko Ngeke, Joseph Lipari, Lisa Lov, Dr. Elia E. Martinez Mercado, Roz McGregor (RIP), Julie McMahon, Fabio Miranda, Andres Moreno, Kisilu Musya, Steve Nelson, Richard, Tatiana Rojas, Saba, Isabella Salton, Aaron Scranton, Tristram Stuart, Mauricio Toro, Henry Tumusiime, Simon Turner, Victor, Marcio Weber, and William.

In addition, Professor Gianmarco Ottaviano and Graham Woodgate, PhD, whom I interviewed in London at the beginning, lent me invaluable insights that helped me decide where to focus my efforts.

For their logistical support, including securing interviews, I would like to thank Sandra Froidevaux, Nimrod Moloto, Wanja Emily, Julie Lunde Lillesæter, and Erika Abiko.

I would also like to thank Mark McPhillips for his excellent English-Japanese interpreting skills; Barhum Nakhlé for being a wonderful host in Bratislava and helping me to secure an important interview, as well as for his English-Arabic translation support; Mrs.

Mieko Ozaki for her warm and kind reception at the Hiroshima Peace Memorial Museum; Samuel Neuenschwander and Dr. Christoph Karlo at Kessel for their support; and Cosette Galindo, PhD, for sharing some of her expert knowledge on the Old Testament and shedding light on queries I had in relation to it.

Finally, I would like to express my gratitude for two important friendships that were strengthened during the course of the journey: Madison Ree Koen, whose cheery spirit and philosophical discussions enlivened the leg from Miami to Boston, together with his continued support of my work; and Daniel Silva, who joined me at the start in London, as well as in Kenya and Uganda during the latter stages to record the interviews and the creative process of writing this book for his documentary film titled *My Little Drop of Honey.*

Introduction

$\mathscr{I}$t was while watching a news report about how we were fast approaching a global population figure of seven billion people that my instincts of curiosity and fear were aroused. Initially, this led me to ask myself how a more populous planet might limit people's access to water, clean air, nutritious food, and other basic things such as education and security. Later, I reflected on how our current way of living was entirely unsustainable and at odds with the natural rhythm of nature. I always knew that we took more than we gave. This was patently obvious. So, too, was the nasty character of industrial agriculture and mining that I had grown accustomed to seeing in documentaries, still photographs, news articles, scientific papers, and books. I was aware of what was happening and complicit in my inaction. So, I chose to do something to rouse myself out of that complicity: I decided to travel the world in search of hope.

My journey across five continents, which took six years to complete, felt similar to walking a tightrope: on one side I witnessed death, destruction, and destitution, while on the other I saw the capacity of the human spirit to overcome seemingly impossible obstacles. I was simultaneously perturbed by the magnitude of the problems I saw and inspired by the courage of those who are committed to overcoming them, often with very limited resources.

The duality, life coexisting with death, hope sprouting from fear, would become a recurring theme during my travels, as would the feeling that humankind and innumerable other species are running out of time. Much of the academic literature, news, and firsthand experience point to a dangerous trajectory that carries most, if not all of life on Earth, toward an increasingly bleak conclusion.

Ever since I was a child, I was encouraged to see the world through the lens of how things change with respect to time. From the age of eight, I played classical violin and created sounds that changed with respect to time. As a chemistry student at university, I spent time in the laboratory and at the computer, analyzing chemical reactions, almost always with respect to time. In banking, be it calculating the price of interest rate derivatives, foreign exchange rates, or government bonds, the idea of the value of money and its relationship with time couldn't have been more fundamental. Time was—and is—an essential part of everyday life.

As a child who received a strict Catholic upbringing, I read the story of Noah's ark several times. The animals boarded this specially made boat, two by two, in order to protect them from the ensuing flood. Today, the temperature, CO_2, methane, and sea levels of our planet are all increasing with respect to time. So, the race against time is to reduce the collateral effects of overexploitation and disrespect of people, trees, insects, and flowers, together with our broader environment, or, in common parlance, to limit the blast radius. We as humankind need to have a period of reflection and act quickly in order to stem the losses that everything and everyone is suffering, before we do, indeed, continue along our increasingly painful and protracted path of self-inflicted extinction.

How would a growing global population that had recently passed the seven billion mark survive on the planet? The academics with whom I spoke in London during the initial stages of my research offered different points of view, but it was during a walk in central London near my old university that I remembered a problem that has persisted: poverty in the United Kingdom.

Indeed, the latter rich-poor phenomenon can no longer be oversimplified through the view of a first world/third world dichotomy; rather, it is a question of a more nuanced Global North and Global South in which poor and underprivileged people are living in cities in the United States and across Europe, while ultra-high net worth individuals and a burgeoning middle class are in several cities across Latin America. Having witnessed these nuances during my extensive travels, it became increasingly clear that socioeconomic imbalances are part of life in the modern era, bringing with them a wide set of challenges.

Explained from a historical perspective in Thomas Piketty's *Capital in the Twenty-First Century*,[1] inheritance and taxation, and the ability to accrue more capital by simply starting with capital, all play their role in creating and exacerbating disparities.

Today, this historical trajectory, which fuels unsustainable growth, is surging the planet toward a painful climax. Tax havens shelter money for the ruling elite, democratically and undemocratically elected government officials, nouveau riche and parvenu alike, and ensure that the rich get richer. Meanwhile, private banking systems can collapse and be bailed out by governments as bankers walk away with huge bonuses and never serve jail time for their reckless and ruthless behavior, yet failure to keep up with mortgage repayments results in becoming homeless. Capitalism has upgraded itself to Capitalism v2.0. Sharper, harder, and more ruthless.

In addition, the feeling of living a precarious reality, with multiple threats including poverty, lack of access to health, education, housing, and the existential dread and anxiety this inherent uncertainty causes, have been issues that I have seen repeatedly almost everywhere I have been. Living on the fringes of society in Atlanta is not a world away from a township in South Africa or a slum in Brazil. People are scared. Some have started to lose hope, and a few have given up altogether, with just a couple who are prepared to fight to the end.

The narrative about a population explosion is neither new nor outlandish. The global population increased by about eighty-two million people between 2018 and 2019, roughly the same as the population of Germany.[2] Each and every new child needs food, water, medicine, clothing, energy, and more, all of which contribute to increases in carbon in the atmosphere.

Just over two hundred years ago, the British scholar Thomas Malthus published an essay about what he believed to be a mathematical link between population and food production, whereby the number of people on the planet would soon outstrip the available food production rates. While the math does not stack up today, he was on the right track: if you have too many people in a finite space with finite resources, once a certain limit has been reached you simply cannot give everyone enough food.

Paul and Anne Ehrlich's book *The Population Explosion*, published in 1990, as a follow-up to Paul Ehrlich's 1968 *The Population Bomb*, explores the critical issues that the authors saw at the time, which include famine, epidemics, pollution, and population growth.[3] Many of the points in the book still hold true more than a quarter of a century later, such as the need to increase people's awareness of environmental issues and the challenges of feeding an increasingly populous world.

When the second book was written, 5.3 billion inhabitants were on Earth.[4] Today, that number is 7.7 billion[5] and, according to the medium variant forecast in 2019 by the United Nations (UN), the global population is expected to reach 9.7 billion by 2050.[6] Put simply, more babies means more mouths to feed, which will increase water stress and deforestation for food production purposes, together with the increased global demands for energy and other resources to satisfy even their most basic needs.

The initial literature survey I carried out revealed some obvious and unsurprising discoveries: we used the equivalent of 1.7 planet Earths[7] in 2014. Although the UN's low and high variant population forecasts have been revised several times during the past seven years,[8, 9] all of them point to significant global population growth.

As I pondered the future, I essentially opened up Pandora's box in my mind from which innumerable questions would emerge, providing a source of debate for me, my interviewees, and those who are close to me. I needed to become selective and pick my battles carefully: carbon, energy, food, water, demographics, ecology, economics, and human rights. A challenge lay ahead of me that seemed deceptively simple: All I needed to do was join the dots and paint the picture of a landscape that was changing rapidly with respect to time.

Significant changes to the planet that humans caused were described in the year 2000 by Paul Crutzen and Eugene Stoermer, winner of the Nobel Prize in chemistry and professor of biology, respectively. They defined the parameters of a geological period called the Anthropocene,[10] beginning at the end of the eighteenth century, to denote the impact of humankind's activities on the planet. In particular, they placed importance on data about CO_2 and methane (CH_4), concentrations that were obtained from ice cores.

After careful consideration, I decided that an endeavor to better understand the links between tangible things such as oil extraction and vehicle pollution, and the intangible, such as people's love and awe for the natural world, would require me to travel extensively over five continents and see things firsthand in order to amplify the voices of the courageous people who are affected by or fighting issues related to climate change. Rather than sit in the comfort of an air-conditioned office, explore ideas, and then pontificate afterward about how others should live their lives, I felt an inner drive to listen carefully, observe humbly, and attempt to explicate what I would discover.

Such an approach, however, created a duality that cut through my journey. The energy I was using to observe the system and simultaneously draw attention to how it could be changed, in both my own life and other people's, concerned me: I was burning fuel and emitting tons of CO_2 trying to raise awareness of the issues facing us and every living thing on the planet, as well as shine more light on some of the solutions that could help us get out of this mess, but doing this would have a detrimental effect through the emission of even more CO_2, thereby creating a paradox in which I would be destroying in an attempt to ameliorate the current situation.

Making the right ecological choices for my own journey was thus fraught with inner conflicts. In my personal life, I rarely fly. I use a bicycle every day and travel by train where possible. Being on a tight budget for my world trip, however, meant that I had to take some ecologically unfriendly options, and pay, whenever possible, a CO_2 offset fee of a few euros to at least calm my conscience and make a tangible contribution toward reducing my carbon footprint.

Each and every plane, bus, and train ticket I bought was done having spent a lot of time evaluating the advantages and drawbacks. I booked every ticket consciously, knowingly, and willingly, always in the hope that the story I would tell, including the narratives of the people I would meet, would justify my journey. The journey itself evolved into four legs: Latin America, the United States and Europe, Africa, and Asia. After each leg, I would return home and take the plethora of notes, assortment of photographs, audio recordings, and video vignettes, then digest them for a while before writing the chapters that are enclosed in this book. In each case, I tried to find the secret center of each place I visited: what did I hear, learn, and feel? What is the most essential part of what I learned?

During my journey, I discovered that the world's energy problems have many solutions, such as solar power in New York City (chapter 6) and wind power in Copenhagen (chapter 8). On the other hand, I also witnessed innumerable examples of excesses and inefficiencies, such as those I saw in the United Arab Emirates (chapter 15). It felt as if each city I visited was at a different stage in a rehab clinic for hydrocarbon addictions: in some of them, citizens' maladies from sedentary lifestyles in sooty environments worthy of inclusion in a Charles Dickens novel were replaced with increased vitality through exercise, cleaner air, and a sense that they were living in cities fit for the twenty-first century, together with the pride that goes with having found innovative solutions.

I spent significant periods of time researching in Mexico City, London, and Berlin, where I was able to make longer-term observations. Following major surgery and during my rehabilitation in Mexico in 2017, I interviewed about fifty urban gardeners in Mexico City over the course of several months, making repeat visits to plots ranging from a few square feet to two hectares. I wanted to get to the bottom of these gardeners' motivations for being connected with nature in an urban jungle, and how this changed them as people.

The interviewees' passion for the natural world encouraged me to convert my own roof terrace into an urban garden where I grew lavender, parsley, tomatoes, beans, corn, and, on one occasion, even managed cotton. Sunflowers also thrived—my favorite type of plant because they grow quickly, are normally quite resilient, and help to attract and give sustenance to butterflies and birds, including different types of hummingbird, as well as bees, all of which came to visit.

I subsequently moved to Berlin where, following a personal tragedy in 2018, I decided to reassess my life by drawing closer to my love for plants and the wildlife they attracted. I dedicated the following twelve months to learning about apiculture (beekeeping) and took photography more seriously, photographing bees around the city. In doing so, I began to learn about how bees forage, work together and reproduce, and gained insight into the vast array of bee species that exist. As my passion for bees grew, I upgraded my camera equipment in order to take macro photographs of them and was able to document more than a dozen different species in Berlin alone, out of about 560 that exist across Germany. At age forty, this was the first time in my life that I had developed an interest in bees: I had never really noticed them before, nor realized their vital importance to pollinating most of the world's crops.

Beauty, poise, and serenity surround the way bees fly and comport themselves, always knowledgeable and precise in their actions. The bees would often stare at the lens, as if posing for the camera. They always knew that I was there, and their graciousness allowed me to photograph them without any issues. I would often spend whole afternoons sitting in front of a beehive in a far-flung part of Berlin or in one of the city's public parks, meditating in their presence. Bees exemplify the fragility of both humankind and everything that lives in our shared environment (chapter 10) and represent all that is good about nature: respect for females through their matriarchal society, their excellent organization skills, intelligence, and beauty. I often observed bees and wondered if the world would be a better place if we behaved more like them.

Much has been made in the media of the prediction that, as of 2019, we have between 1.5 and 11 years to fix the climate issues. It is a widely accepted fact that the Arctic and Antarctic are melting at unprecedented rates, that sea levels are rising, and that species are being lost in epic proportions that are difficult to imagine. The language has changed recently. The concept of "climate change" is now known as the "climate emergency" thanks to the young Swedish activist Greta Thunberg, but even that does not really describe the collection of symptoms accurately.

The myriad of maladies, which will not discriminate between classes, races, or genders as they gather more momentum, are best described by a term called the Sixth Extinction, a phrase that is on the fringes of public discourse and yet accurately summarizes what I saw with my own eyes in numerous places. One example of this would be Mexico, where I only saw two owls and a few fish swim in a lake inside the Naha jungle in the southeast of the country (chapter 1). I could not see or hear monkeys, jaguars, or any large mammals that one normally would expect to see in their natural habitat. Upon leaving the jungle, I saw out-and-out destruction: tropical trees had been cut down, a process that released CO_2 to make way for cattle, which belch methane,[11] a gas that is more potent than CO_2 and that produces CO_2 through flatulence. Meanwhile, men carrying backpacks were spraying crops with pesticides to ensure higher crop yields, thus driving the final nail in the coffin for this delicate environment, where toxins were starting to accumulate.

The images from the jungle would repeat themselves many times during my journey across the world and fill me with a sense of angst and despair at the scale of deforestation, which, in Latin America, especially in Brazil, parts of sub-Saharan Africa, and southeast Asia, are taking the planet to an irreversible tipping point. Just before the publication of this book, Arctic forests, the Amazon, and Australia were on fire, with plumes of smoke so intense that they were sometimes visible from space.

But extinction isn't limited just to species affected by the destruction of their forests and other habitats and replacement of biodiverse landscapes with monocrops. The trend I see is one of ubiquity, uniformity, and blandness at cultural, ecological, and spiritual levels. God, the human spirit, colorful insects, birds, and languages have been reduced to a commodified thing that can be bought and sold in an exchange using an app. That's how extinction will really transcend.

Another challenge is that human rights and human dignity are at risk in many industries, from farming (chapter 5) to mining (chapter 14), due to precarious working environments and a lack of law enforcement. I began to consider ethics as lying at the heart of the problematics of extraction, farming, distribution, and sale of food and raw materials. This, in turn, encouraged me to increase my sensitivity and openness to better empathize with people who have been heavily affected by the heartless, ruthless juggernaut of unbridled capitalism.

Mercifully, I saw opposite (but unequal) forces at work in attempts to arrest, or at least curtail, a range of malevolent forces. In Kenya (chapter 12), I met Kisilu Musya, a farmer, video diarist, and climate activist who is working on numerous projects to protect his farm and local community from the effects of floods and droughts through ecological farming methods and water management. Included is a newly constructed dam that serves as a reservoir, a buffer against the effects of unpredictable rainfall.

In central Brazil (chapter 3), I visited the Earth Institute (*Instituto Terra*), which is set in a valley and part of the Brazilian Atlantic forest that had been reborn following an intervention by renowned photographer Sebastião Salgado and his wife, Lélia Wanick Salgado. They had transformed the previously barren valley into a lush haven to which wildlife returned. It was there that the seed of love, respect, and awe I have for trees was sown.

Some other changes crept in during my journey: when I learned about deforestation in Indonesia and the plight of orangutans, I began to eliminate products that I knew contained palm oil; I stopped buying fast fashion clothes and either bought secondhand clothes, or, in some cases, bought new fair trade organic clothes, which are expensive but worth every penny. I stopped using plastic bags entirely. I started buying local where possible, and drastically reduced my meat and poultry intake, among other things. I often discuss these topics with people who are close to me. Each and every choice can make a difference.

Beyond these everyday decisions, I gave up a comfortable life nine years ago to pursue my dream of becoming a writer. Rather than focus on fiction, I decided to spend time in the field and listen to others who spoke their truths to me, then weave those narratives into a tapestry that would show how connected we all are. My motivation was to understand the links between peace, education, and the art of making music, which resulted in a book and film project called *Music and Coexistence*. Toward the end of that manuscript, in May 2014, the idea for this book was born.

Over the course of the subsequent journey, I fell in love with the natural world and, in particular, bees. I see the parallels between humanity and innumerable other species to the degree that I no longer use the us/them dichotomy, preferring only to employ the pronoun "us" because we only have one shared destiny. Even though future events will determine the outcome of this highly turbulent chapter of earth's history, it's clear that the climate crisis is the defining force that binds all living things, because the ecosystem that provides the framework for life as we know it is under high levels of stress.

Although humankind and innumerable other species are on a dangerous trajectory, I took comfort from meeting so many brave, wise, and committed people during my travels. History and literature remind us that plot twists are an inevitable part of character development and dramatically change the outcome that predictions foresee. With this book, it is my hope that the actions, voices, and sentiments of the remarkable people I met who are striving to make the planet a more livable, benevolent, caring, compassionate, biologically diverse, fertile, and beautiful place for everything and everyone may be shared with others. Ideally, the chapters should be read in order because, broadly speaking, the narrative has been arranged in chronological order. That being said, it is entirely possible to review parts of the book at different points in time without following a specific order.

I

LATIN AMERICA

From the Jungles of Mexico
to the Skyscrapers of Panama

DEPARTURE

I flicked open my notebook, ready to begin pouring out my observations, thoughts, and feelings over the next twelve thousand miles across a continent I knew and loved. The traffic began to clear in the east of Mexico City, the bus gathering pace, the overcrowded shantytowns at the city limits, similar to those I would see throughout the continent, beginning to emerge, sitting precariously on hilltops where they erased the trees that once stood there. The gray toxic smoke that, during my time in this most dense city, had polluted my lungs and made me sneeze and sniffle constantly, started to disappear, and I felt like my voyage finally had begun.

During the overnight bus ride to San Cristóbal, a picturesque colonial town in southern Mexico, I pondered a fundamental question I would visit repeatedly during my trip: how on Earth will we provide for a growing population increasing with respect to time? According to figures published by the United Nations in 2019, the world's population will increase by two billion within the next thirty years (medium variant).[1]

Note two important things: First, the UN population forecasts contain a great degree of variation, with the maximum and minimum figures for 2100 being 15.6 billion and 7.3 billion,[2] respectively. Put simply, this variation is due to the UN's, as well as any model's, inherent uncertainty and greater unpredictability as the time horizon is stretched. Second, this growth is not expected to be uniform; the population in some regions, such as in Africa, is predicted to expand, while populations in certain places, such as Cuba and Ukraine, are expected to contract.[3]

So, we *know* that more people will be on planet Earth, which undoubtedly will strain existing resources more and create more pollution. However, rather than project into the future, an important part of my search had to do with the existing situation: where is humanity today? Given our massively growing population, what is our relationship with the environment? Are we headed toward disaster, or is this just an exaggeration by the so-called liberal left?

Predictions, by nature, can be unreliable. Even great minds may make inaccurate assertions, such as the British scholar Thomas Malthus, who stated in *An Essay on the Principle of Population* in 1798 that population increases in a geometric ratio, whereas subsistence increases in a mathematical one, meaning that the rate of population growth would exceed food production and, therefore, have dire consequences.[4] Although the math was incorrect, he did raise a valid point: a moment comes at which consumption outstrips supply.

In a recent extension of Malthus's ideas, Paul and Anne Ehrlich warn, "We shouldn't delude ourselves: the population explosion will come to an end before very long. The only remaining question is whether it will be halted through the humane method of birth control, or by wiping out the surplus."[5]

Taking this idea a step further still is the term "sixth mass extinction," which has numerous advocates behind it, including Elizabeth Kolbert[6] and Noam Chomsky.[7] Essentially, this is a doomsday scenario in which life as we know it ceases to exist because the pillars that support it have collapsed due to humankind's actions over the past two hundred years. The latter time period coincides with the industrial revolution and a marked increase in the global population: the planet's population was fewer than a billion people in the year 1800,[8] compared with nearly eight billion today. In fact, this extinction has already started. We've seen massive species loss in recent years, and, as I would discover on my journey, ecosystems are changing rapidly due to human activity.

According to the Alvarez hypothesis, dinosaurs were made extinct by a meteorite[9] that caused the climate to change through the dust produced by the impact.[10] Those reptiles had neither the means to foresee nor the ability to protect themselves from this disaster. Notwithstanding the fact that we have the greatest number of refugees in the history of humankind (twenty-six million, approximately the population of Australia),[11] as well as increasing levels of desertification that could displace fifty million people in the next ten years, putting even more lives at risk,[12] together with other existential threats, there are many sources from which it can be possible to envisage a way to overcome the sixth mass extinction, which has already begun.

Recent decades have seen innumerable advances in science and technology, including the internet, which, when combined with the human will to survive, would mean that my journey would encompass both problems and solutions through which to observe the world with greater balance and hope. I wondered whether we have enough time to mitigate the damaging effects that have already been caused. I was—and still am—driven by hope, rather than paralyzed by fear, because I believe in the intrinsic goodness of the human spirit, even if, as I would discover, we have numerous reasons to feel terrified.

WHERE ARE ALL OF THE ANIMALS?

I met with a village elder, a quiet and unassuming man who is a member of one of eighty-six families that live in Naha, a remote jungle community about four hours from San Cristóbal. The village is home to the Lacandonan people who speak Maya, a language that is common both in that region of Mexico and in Guatemala. I looked above his head and saw a landscape painting of his village, depicting a traditional Mayan religious ceremony. Everybody—men and women—was dressed in white. I also noticed the late Chan Kin "Viejo" with long, dark hair portrayed in the painting; earlier at the ecolodge I had read lines quoting him:

> Hach Akium, God of all Gods, created heaven and the jungle. He sowed stars in the heaven and trees in the jungle: the stars have only one root . . . which is why when a tree falls, so does a star.

This really touched me as I often look up at the sky on a clear night and marvel at the beauty of the stars. Imagining them falling out of the sky compelled me to ask the elder to go into more detail about the words of Chan Kin "Viejo": "If we want to cut down a tree, we first have to ask permission from God, because if we don't, when the tree is cut down, a star will also fall, therefore, we have to look after it [the jungle], because [it] is God who sowed the trees." The elder believes that the jungle asks for oxygen, wind, and rain, and that his community has to ask God to harvest the *milpa* (corn, beans, and zucchini), a staple in the Mayan and Mexican diet. After each cycle, they burn the *milpa*—a common practice in the region—and, later, sow seeds, after which then they have a family gathering so that God blesses the new harvest.

"We have to continue moving forward, we have to fight, the children are growing up. We have to teach them so that they can tackle [these challenges] . . . if I die, then my children and grandchildren know how to look after the jungle," said the elder, explaining further: "We don't want to use fertilizers or chemicals, we want it to be natural . . . that they [the children] continue sowing corn and beans, and that they don't change [the methods] and that things stay the same."

I moved the conversation to sustainable tourism, an area in which the community has made an investment with government support, including a rather fancy hotel with a thatched roof to make it blend in with the local environment. The community is trying to tap into the premium tourist market, which can not only increase revenue, but also reduce dependency on farming and the underlying risks wrought by a failed harvest.

The community offers visitors tours of the *milpas* and a number of traditional ceremonies, including a *limpia* (cleansing): "God can heal everything. He hears everything." This Mayan ceremony, which I've witnessed a number of times across Mexico, normally is accompanied with drums and other percussion instruments. During the ceremony, the recipient inhales smoke, which, it is believed, will cleanse him and ensure a safe passage or assuage his pain.

All of the men in the village wear the white tunic and are encouraged to wear their hair long because "that's what the tourists want to see," helping to support the illusion that this is in some way a living museum where they can visit pristine lagoons and a nearby pit in the middle of the jungle approximately one hundred meters deep that is full of trees. I was skeptical.

At the end of our meeting, I asked the elder if I could photograph him, but he politely declined, and I recalled that as a guest, he was under no obligation to allow me to do so. I thanked him for his time and walked back toward the lake in the dark. The night was overcast, which meant I couldn't see the stars as I'd wished, but I could see the neat, humble, and spotlessly clean homes, the community clinic (part of IMSS, the Mexican National Health Service), and the police station, which had a massive pickup vehicle outside.

The elder was cordial but guarded, and I sensed that perhaps the place was too good to be true. I wondered whether other forces might be at work. Having traveled to eighteen of Mexico's thirty-one states during more than a decade of living in the country, I was left scratching my head. I hadn't seen anything like that before. Could this village survive on tourism alone? Why did the Mexican government have such a big presence there when, in innumerable other villages I had been to across the country, they seemed to be almost absent? I never was able to resolve that tension, even after I completed my journey, but it encouraged me to question the role of the state both in Mexico and other Latin American countries as I delved more deeply into the region's complexities.

Apart from a rowing trip on my own around Lake Naha, I decided to spend my penultimate day sitting on the jetty, made of wooden planks strapped to plastic drums to keep them afloat, and photographed trees on the other side of the lake, practicing time-lapse photography and video. By using the movement of the clouds and their reflections against the stillness of the lake to cause some visual drama, I was able to record a wonderful video with the most impressive soundtrack, worthy of any chillout album, with bleeps, cracks, hums, and whistles of insects interspersed with the occasional birdsong, which caused a sedative effect.

The humidity was a killer, but an attractive force kept me there—such was the undeniable beauty that I was witnessing. In the distance I saw a resident of Naha, a young man, rowing slowly back to the jetty. He took a seat next to me, then signaled me to look down to the clear water, and in a cautious attempt to break the ice told me, "there are sardines and crappie here," the latter being a tropical fish common in Mexico, darting around the reeds and lilies.

The young man pointed into the distance, telling me of a neighboring village where there used to be jungle; it is used for livestock farming, not something the Lacandonans do. He went on to observe, "They didn't dedicate themselves to preserve [their environment] . . . Nature provides us with clean air, we never get sick because we look after the jungle." According to him, animals such as deer, boar, tepezcuintle (lowland paca), and jaguar live in the jungle; therefore, in order to preserve the species, they don't hunt them. Oddly, I hadn't seen any animals during my time there even though I was in the middle of the jungle, far away from any roads and any noises that might scare them, which made me question the accuracy of his observation.

I meditated further on what I had seen—the fragile ecosystem and small community that could become extinct or simply absorbed into mass popular culture. I also pondered the big picture—where Mexico is headed—given that it is one of the world's seventeen megadiverse countries that contains between 10 and 12 percent of the world's species, even though it only represents 1.4 percent of the surface area.[13]

On my final day, a fellow guest at the eco lodge, Alberto Martinez, a photographer and owner of an environmental conservation NGO, offered me a ride back to San Cristóbal over rough terrain. After a few stops to take photos around nearby Lake Ocotalito with the 6 a.m. mist forming a veil over the tranquil lake, I got straight to business. The question had been niggling at me for a while: why hadn't I seen or heard any mammals, including jaguars and monkeys? Surely they should be abundant in a habitat such as this? I asked him whether the monkeys have been hunted.

He replied: "Yes, or, they're hidden, or far away."[14]

After three trips to Naha, he had yet to see one. I asked him whether he'd seen deer.

"Of course," he said, "they should be here, but we haven't seen deer during our visits."[15]

He added that neither he nor his friends had seen mammals in the jungle. Were the bigger animals hiding somewhere, or had they been hunted to extinction? That birds were there meant that the forest could support life. But birds are notoriously hard to catch, and small ones are not worth the effort, putting hunting high on my list of possible reasons for their absence.

We continued our cautious drive over the gravel road with massive ridges, which soon turned into a highway, something that concerned Alberto.

"A highway is a communication medium, which will allow people to visit Naha, but if the local people don't have awareness about the driving speed and that they can kill the fauna that crosses it, it will be negative,"[16] he said, referring to an owl, a female about the size of my hand, that we had come across earlier.

As we left Naha, I saw the bubble that the village elder spoke about as clearly as I ever will: once we had reached the village limits, the dense concentration of trees of the jungle were replaced by large fields with cows that looked as if they'd be more at home on an alpine mountain than in the middle of what used to be a jungle (although, of course, much European livestock in Europe grazes in what used to be forest).

Our one-ton friends grazed and mooed, releasing CO_2 and methane (CH_4)—a double whammy of greenhouse gases, which became a triple whammy with the reduced CO_2 absorption given the dearth of trees. I stopped to take a few photos of the cows grazing in a field with Naha in the background. This was both concerning and surreal; not only were the cows not in their natural habitat, but the habitat in which they now lived had been completely different, having converted from jungle to pasture in just a few days' work.

We passed through villages and decimated pine forests. The scars of red earth looked wounded; the stumps of amputated chargrilled trees oozed pain into the sorry soil. Abattoirs were frequent, too; the final stop in the cows' lives. How little did they know about their fate. Meanwhile, men with white plastic backpacks were happily spraying the fields with pesticides to kill pests and, while they were at it, were eliminating beneficial insects, in doing so causing more damage to the environment and contributing to the sixth mass extinction of the species.

The ecosystems changed haphazardly. Alberto explaining each change as we passed through some stunning oak and pine forests. As the Jeep wound through a tight bend, I could see mist at the mountaintop. I recalled how trees at the top survived because they weren't easy pickings. Then we slowed because a truck ahead of us carrying tons of pine logs had stopped. I quickly reached for my camera and took a few shots, then put it away as we carefully overtook this stationary forty-foot hearse where the recently slain ecosystems lay motionless.

Once we returned to the city limits, my mobile internet was working again and, a few swipes later, I could see the satellite images that I'd been craving since I arrived in Naha. The village elder was right: they live in a bubble, the last bastion of jungle at the very edge of a vast, diverse, and beautiful country. Now it has numerous plots of adjacent land no longer dark green on the maps as they should be; consequently, the jungle is unable to support the wildlife that previously inhabited it.

Deforestation is a dark, shady business. A few months after I left Mexico, the country had another high-profile murder of an environmental activist. This time, it was Isidro Baldenegro, shot in January 2017. He had previously won the 2005 Goldman Environmental Prize[17] for protesting against illegal logging in the northern Sierra Madre region, which is home to the indigenous Tarahumara people to which he belonged.

After spending a night in the quaint and beautiful San Cristóbal, I took a bus toward Guatemala. About a mile before we reached the border, traffic had ground to a halt and we all had to get off of the bus and walk beneath the fiendishly hot spring sun. Local farmers had created a roadblock because they didn't have any water in their community. The European tourists who were on the bus didn't seem to understand the scale of the issue and began to moan.

One passenger shouted at the driver, "What's going on? We paid for our tickets to Guatemala, and now we have to walk in this heat?"

Pine logs aboard a forty-foot trailer in Chiapas, Mexico. Photograph by author, 2016

The whole event probably seemed like an inconvenience to their precious backpacking trip across Latin America because, in Europe and other parts of the Global North, this seldom happens. Delays are normal, almost expected, in the Global South.

Once I reached the hotel in Guatemala five hours later, I did an online search for Cuenca de Grijalva, where the protests about the water shortage had taken place, and neighboring towns, for breaking news. I tried again the next day, and the day after that, and found nothing. Their cries for help fell on deaf ears. I had left my dear and cherished Mexico and had begun to see it in a new light. The country that I love, endowed with warm people, a rich history, natural splendors, and irreplaceable beauties, was going through massive economic and social change. Protests seemed to have become almost ubiquitous in Mexico City. Although the cause for each protest was different, ranging from women's rights, femicides, water shortages, to electoral fraud and freedom of speech, the root cause was the same: injustices caused by polarization and systematic abuse of power.

ESPRESSO IN GUATEMALA

Following inspection of the unripe green beans, the chief taster Umberto Tahuite, a man with more than thirty years' experience, continued to prepare the coffee. After roasting the beans, he prepared five cups of coffee, then followed with a rapid sequence of slurps, sloshes, squints, and spits. After writing down some notes, he decided on the future of the collective that had submitted its coffee for review: "It's good. It can be exported."

But not all coffee submitted to ACODIHUE,[18] an organization that collects coffee from farmers across the Guatemalan state of Huehuetenango, can be exported. The mountainous region in the northwest of Guatemala has fertile soil. That, together with an apt microclimate, creates the conditions to grow rich and flavorsome coffee; nevertheless, sometimes the taste simply isn't good enough to meet foreign buyers' requirements, even though it can satisfy local needs. In some cases, the unloved beans are fermented, yet still produce decent-tasting coffee, in which case they can be offered to the domestic market. If a buyer can be found, the beans can be resold on their own, or blended to make a low-grade, cheap coffee.

Coffee provides a major source of hard-currency income in Guatemala; sales from the 2018–2019 harvest totaled US$660 million, representing 7.3 percent of the nation's exports and positioning the country as the ninth biggest coffee exporter in the world.[19] This industry provides employment for thousands of families eager to sell their best beans in foreign markets. Ultimately, this means that the pricing of such a valuable commodity is crucial in order for the farmers to get as much money as possible.

I spent a few days shadowing ACODIHUE's export manager, Felix Camposeco, who explained the coffee pricing model,[20] which seemed logical: if a producer can demonstrate that he follows the fair trade model based on social and environmental sustainability, and meets organic standards, then an organic premium is added to the futures price set in New York, along with a country premium,[21] to give a final offer price to the producer.

An essential part of securing a good price for the producer and, therefore, ACODIHUE, is to purchase coffee futures contracts to protect both parties from adverse price movements in the coffee market. The use of futures contracts and options told me how far reaching and intertwined markets have become. While visiting producers on a mountain a few hours away, close to Todo Santos, I recalled Davos, Switzerland, where the World Economic Forum is held in which bankers, politicians, and the world's elite decide the fate of the world.

The advance of investment banking, brokerages, asset managers and hedge funds has meant the proliferation of electronic trading, algorithmic trading, and arbitrage trading, among others, which encouraged more players into the market to fulfill their ambition of wealth creation. That is, it brought players who are not interested in the physical delivery of gold, coffee, oil, or whatever underlying asset that they are trading, but, rather, just want to make more money.

This is clearly exemplified in advertisements for brokers on social media: the investor uses a mobile app to enter a few trades, make quick money, and have a good life—to which everyone is entitled—because he made a profit on the markets, and it's time to relax and enjoy the fruits of his labor.

What these ads don't show is that many farmers—pretty much every farmer I met on this journey—barely make ends meet. They walk a precarious path with no safety net, without any social security system to speak of, making them sacrificial pawns in the financial casino. To put this in context: each and every day, investors trade futures and options on commodities and are never accountable, yet if the producers make any small error along the way, they stand to lose everything.

Seeing this bigger picture, I asked myself how the way in which the market is used could be a problem. The market, like any tool, has more to do with intent and use than with anatomical and physiological descriptions of its form and function. It can facilitate business transactions and help spread wealth. Our destinies overlap and are intrinsically linked: it's how we recognize the other, and the accountability and responsibility that come with this recognition, that will determine how nearly eight billion people will be able to get a fair piece of the pie.

ACODIHUE has very few options other than to use the financial markets to help reduce its financial risks if, for example, coffee prices drop sharply. In itself, this reveals a paradox: most trades in any market, be it foreign exchange, equity, commodities, or any other that comes to mind, are largely speculative, and participants play a game of pass the parcel, with the hope that they make a profit and don't get burned (the phrase "for every winner there's a loser" was one of the first I heard on the trading floor in London). However, it's precisely the market that provides the insurance policy that Felix spoke of, and simultaneously contributes to the topsy-turviness of prices, also known as volatility, which is where technical traders and savvy investors make money. "Buy low, sell high" is the mantra that transcends space, time, and financial instrument type.

It's hard to argue that a coffee trader who lives and works in a financial center should earn hundreds of thousands of dollars a year when the person who toils in the field receives a pittance. This isn't Marxist discourse—it's common sense. If we continue to exacerbate inequalities around the world, then we will continue to see waves of economic migration, which has happened throughout history and will continue to do so for the foreseeable future. In 2019, Guatemalan expatriates sent home more than US$10 billion in remittances,[22] which is approximately fifteen times more than the entire Guatemalan coffee export market.[23]

I'll return to the theme of remittances in chapter 5, when I detail the completion of my trip across Latin America.

Fair trade's potential to change lives, which includes producers of bananas, cocoa, and tea, to wine and textiles in the Global South, would, in principle, be a good way to make a difference when buying a product produced thousands of miles away. By offering a premium to producers, fair trade gives them a much-needed economic boost and with it the possibility of improving health care, environment, and education. Nonetheless, fair trade is not a panacea, because the problems that farmers face are far too complex, not to mention the carbon footprint related to goods as they move around the world.

The reality in Guatemala, and to differing degrees in other countries in central America, Africa, and Asia today, is extreme violence and high emigration rates, combined with high corruption indexes and a lack of social mobility, justice, and democracy, which come together to paint a bleak, complex picture. Therefore, in order for fair trade to have significant impact at a national level, these other factors would need to be improved to make an enduring mark on producers' lives, along with ramping up production and exports to replace non-fair trade products. The downside would be the carbon footprint that increased trade causes (this will be explored further later in this chapter).

In the mountains of western Guatemala, close to the border with Mexico, the four-wheel-drive truck cautiously navigated steep canyons, waterfalls, and rivers, as well as man-made smuggling routes for people and contraband goods. Three backbreaking hours later, I arrived in San Antonio, a small and tranquil village, where I would rest before going to an event the following day that was organized by ACODIHUE, with support from Australian Aid.

This event, an organic fertilizer workshop at the local primary school, was held by Felix and his colleague Carlos, a specialist in organic farming methods, with support from a translator of a Mayan dialect spoken in this region, who gave regular summaries of the talks to the local farmers who sat on tiny chairs in the classroom.

Carlos presented a practical session for farmers on what organic fertilizers were and how to make them. At times, I felt a disconnect between the scientific nature of the talk of "organic nutrients" and "copper sulfate" and the farmers' reality, which is simple and modest: many of them speak only a few words of Spanish and have no formal education.

The local coffee farmers were offered free food as an incentive to be there, and a consensus between many of those I spoke to afterward was summarized by, "I will sell to the highest bidder." ACODIHUE essentially facilitates and reconciles two sets of requirements: the financial rewards offered by foreign buyers interested in acquiring fair trade and organic coffee, which they can sell to consumers at a profit, and the farmers' need for survival.

After the workshop, I asked Carlos about water shortages, to which he replied, "It doesn't rain, which means that your harvest is either depleted or is reduced."[24] Both temperatures and demands for water have risen, and water sources have been depleted, which has meant that the ground has become drier, leading to coffee plants absorbing even less water and putting pressure on coffee yields, or potentially resulting in outright crop failure.

Such environmental risks pose an existential threat to the communities that live off the land, making them even more vulnerable to climate change. Just as with the communities that protested the drought in the Cuenca de Grijalva area as I left Mexico, only a few miles across the border, one thing became clear: the current use of land, which is unsustainable, has reached a breaking point.

NICARAGUAN RUM

My instincts told me that something wasn't quite right when I saw the motorcycle coming toward me. The rider stopped in front of me, facing the oncoming traffic, and motioned me to come closer to him. I started screaming—the human response I'd imagine many people to have in such a situation.

"Please don't rob me!" I shouted in Spanish.

No way could I run away. I was fully laden: US$1,000 in cash, passport, two credit cards, camera, lenses, tripod, laptop, and, of course, my luggage. Luckily, a car stopped.

The driver got out and shouted to me, "Hey, bro, what's goin' on?"

As more people gathered to watch the spectacle, the rider, who in Argentina would have been called a *motochorro* (quite literally, "motorcycle thief," who is likely to be armed), realizing that he had missed his opportunity, flicked his black visor down so as not to be recognized, and sped off. A teenager suggested that I should start running to avoid the man catching up with me later. He may have gone around the block, only to return from another direction. If he were to succeed, a year of preparations would be down the drain on only the third week of my world trip. Carrying about forty pounds of gear, I hobbled as fast as I could in the opposite direction.

I arrived at the hotel an absolute wreck. The disagreeably damp afternoon in Managua, Nicaragua, seemed unlike any other. I was drenched from the humidity and doubly soaked in my own sweat, making the humidity feel like 200 percent. I was hyperventilating, and I had a panic attack. I started to inhale through my nose and out of my mouth several times. At my request, the owner, a sweet Spanish man in his early seventies, poured me a small glass of white rum to calm me.

While I'd never claim alcohol as a cure for any malady, I can say that that day a glass did exactly what was needed: fast acting, it helped take the edge off, without any apparent side effects. I felt better within an hour. His wife, a kind Nicaraguan woman, prepared some pulled pork and rice for dinner. I felt that their warmth and hospitality, not to mention their genuine concern for my well-being, more than compensated for a horrendous experience in their country. I recognized that although the instability I had seen in Nicaragua and other Latin American countries unfortunately is common, the vast majority of people are inherently well intentioned and must improvise to counteract those malevolent forces.

During the next few days I spent most of my time indoors, recovering my strength after the journey from northern Guatemala, which involved two overnight bus trips sitting on what felt like a cast-iron seat with wafer-thin padding. That was the easy part. The hard part was passing through two of the most violent countries on earth, El Salvador and Honduras, knowing that the bus could be stopped at any time and we could be robbed, or worse. I had been to both countries six years earlier, during my banking days. Back then, I had been scared when I went to a restaurant in San Salvador where a security guard with a pump-action shotgun stood outside; and in Tegucigalpa, the capital of

Honduras, where I was told that I could only go to the mall across the street but not walk around the city, even in daylight.

After some much-needed rest and relaxation in beautiful surroundings with my wonderful hosts, I began to secure key interviews farther down the road in Colombia and Brazil (detailed in chapters 2 and 3, respectively).

A key theme I wanted to explore was how to reduce or reverse the dangerous levels of carbon emissions that were cooking the planet. In order to gain a better understanding, I began to review the Kyoto Protocol, signed by 192 parties.[25] First published in 1998, it identified six gases as contributors to the greenhouse effect,[26] or the storage of the sun's radiation in the Earth's atmosphere that causes an increase in temperature: carbon dioxide (CO_2), methane (CH_4), nitrous oxide (N_2O), hydrofluorocarbons (HFCs), perfluorocarbons (PFCs), and sulfur hexafluoride (SF_6).

Looking at global warming and the associated climate crisis through a wider optic, they can be held responsible for a multitude of issues that I would see throughout my journey: from bees that had to drink from a water fountain during an exceptionally hot summer in Berlin (chapter 10), to low water levels in the Kariba dam that straddles Zimbabwe and Zambia (chapter 13), causing power outages in both countries; or the unpredictable harvests and weather patterns for which Kisilu Musya, a farmer and activist in Kenya, would have to find solutions (chapter 12).

During my final days in Nicaragua, I began to question my own habits and activities. I thought perhaps I should at least start to think about the most ecologically friendly mode of transportation possible, with the caveat being my safety. Through this, my aim would be to reduce my CO_2 emissions rather than live in a bubble and fly everywhere as I had done in the past, giving no consideration to the effects of my actions. I prepared a list of the modes of transportation that I had used, and I tried, as far as possible, to calculate my CO_2 emissions. It wasn't as easy as I had first thought.

Two key issues arose: to begin with, I wasn't aware of a real-time CO_2 calculator for any mode of transportation, which meant that I would need to use existing online sources for emissions figures created by automobile, bus, and aircraft manufacturers in cases where the model/engine type was known. In cases where this wasn't clear, such as riding a bus that didn't have a known chassis or engine details, I would try to find the details of a bus that had a similar engine, size, and weight with known emissions levels.

The second issue, which is even more complex, was to work out how much CO_2 was produced in creating the fuel in the first place. To my surprise, unlike food, which in almost every country I've been to is labeled according to calories, fat, carbohydrates, protein, and so forth, nothing comes even close for fuel emissions or the production and transportation of fuel from oilfield to gas station, despite its massive impact on our health and the environment. Both extraction and transportation emissions would need to be combined to arrive at the true value of the CO_2 outputs.

While policy makers, governments, scientists, and industry continue to battle out the future direction of fuel exploration, extraction, and end use, I felt some empowerment as I began to make informed, albeit limited, decisions about my carbon footprint. I will return to these questions and some of the outcomes in chapter 9, as I recount reaching the end of the European leg of my journey. Back in Nicaragua, a stark reminder of the effects of global warming came with the frequent power outages, which meant that I would baste in my own sweat in the desperately humid city of Managua.

CANALS AND PAPERS

I was quite surprised to arrive in Panama City a few months after the release of the Panama Papers, and sensed that I was in the Dubai of Latin America: high-rise buildings, fancy cars, yachts, even a Donald Trump building. The dichotomous rich-poor divide, which has been an essential feature of empires, countries, and cities for millennia, can be seen clearly today in Panama City and other major cities in Latin America and the rest of the world. As an Oxfam paper titled *An Economy for the 1%* asserts,

> The global inequality crisis is reaching new extremes. The richest 1% now have more wealth than the rest of the world combined. Power and privilege is being used to skew the economic system to increase the gap between the richest and the rest. A global network of tax havens further enables the richest individuals to hide $7.6 tril-

lion. The fight against poverty will not be won until the inequality crisis is tackled.[27]

The Oxfam research demonstrates that the gap is widening. Sixty-two people had the same wealth as the bottom 3.2 billion (almost half the world's population) in 2015 compared with 388 in 2010, and the bottom half lost 41 percent of their wealth during the same period.[28] The report also points out the effects of a tax shortfall, implying tax increases for smaller businesses and less affluent echelons of society, or cutbacks, both of which occur to the detriment of the people who need the services the most, further exacerbating inequalities.

The Panama Papers scandal in 2016 underlines the global and systemic nature of inequality. Some 11.5 million documents[29] from Mossack Fonseca, a Panama-based law firm, were leaked by nearly four hundred journalists[30] in what would become the biggest data leak in history.[31] In the days, weeks, and months that followed, I read about the resignation of the Icelandic prime minister, Sigmundur David Gunnlaugsson, the compliance actions against one hundred Australians,[32] a sixty-five-member panel undertaking a ten-month enquiry at the European parliament,[33] and the Danish authorities even spending 805,000 euros on an anonymous source for data connected with the papers.[34] The list of examples continued.

What the disclosure of the Panama Papers achieved is remarkable, and it marks an important step toward fighting the inequality Oxfam described. A team of journalists joined together in a collaborative task that spanned languages, cultures, time zones, and additional obstacles to provide both the general public and government bodies with greater transparency relating to offshore schemes—to bring about change. It demonstrates that knowledge is, indeed, power.

The parallels between the Panama Papers and previous scandals that rocked the financial world seemed stark: First, two separate foreign exchange price-fixing schemes[35] (the first between 2009 and 2012, the second between 2007 and 2013), where traders from different banks colluded to create artificial exchange rates, had massive effects on the valuations of financial products that relied on rates they published. In the London Interbank Offer Rate (LIBOR interest rate) scandal in 2012,[36] traders from a number of banks made either inflated or deflated submissions of their interest rates to make profits and, in the process, defrauded the rest of the system, including bond and interest rate derivatives markets, and of course, the public at large. Another major event happened at Enron, the energy company, which kept debts off its balance sheets, meaning that investors who had bought artificially inflated stocks took a financial hit, and, of course, staff lost their jobs.[37]

What makes the Panama Papers somewhat unusual, and indeed troublesome, is the scale and scope of the practice of offshoring and tax avoidance that it highlighted: the scandal revealed the involvement of politicians, their associates and family members, as well as business leaders, sports people, and other high profile figures, creating an eclectic group of people from practically every country in the world who had common objectives. The then-prime minister of the United Kingdom, David Cameron, said: "Frankly some of these schemes where people are parking huge amounts of money offshore and taking loans back to just minimize their tax rates, it is not morally acceptable."[38]

Perhaps feelings of superiority and entitlement are common features among people who don't wish to pay their fair share of tax. They lead to the assumption that the rules that apply to the masses don't apply to them because they are rich and powerful, in turn meaning that loopholes can and should be exploited and, if necessary, laws broken and integrity used as collateral in pursuit of accruing more wealth in order to move up the *Forbes* list.

Interestingly, as I reflected on this in the epicenter of the scandal itself, I didn't see beggars on the streets, marches, riots, or anything that would indicate any trouble at all. Panama City seemed eerily normal, which ignited my curiosity further: people were jogging, families were doing their weekly grocery shopping while wearing expensive jewelry, the highways were filled with luxury cars. At times, I could have been mistaken for thinking that I was in the Global North. How could this piece of paradise sustain itself economically? Was money laundering and tax evasion an open secret, just part and parcel of doing business in Panama?

After several unsuccessful attempts through online searches to secure firsthand accounts of what happened, I decided to visit the Panama Canal. The cargo ships passing through were enormous blocks of steel, hundreds of yards long, displacing hundreds of thousands of gallons of fresh water, assisted by four trains, one at each corner slowly, yet efficiently, dragging the huge vessels into the canal to be either filled or drained, depending on which direction they were headed.

The global shipping industry is an almost invisible, silent set of vessels that transport goods cheaply around the world. But this inexpensive transportation comes at a price: shipping contributes nearly 3 percent of global greenhouse gas (GHG) emissions. Also, the heavy fuel that comes from the distillation of crude oil contains high levels of sulfur[39] that, when combined with the plumes of dirty exhaust fumes filled with particulate matter, have a negative impact on human health. Mercifully, plans were to cap the amount of sulfur in shipping fuel effective January 1, 2020.

On the horizon, I saw dozens of ships patiently queuing up in the Pacific Ocean, waiting for this shortcut to the East Coast, Europe, or elsewhere. The future might look different. According to an article published by the Yale School of Forestry & Environmental Studies (now the Yale School of the Environment), increased global warming will create further melting of the Arctic region,[40] which will provide shipping lanes that will reduce the distances, times, and costs of goods moving between China/Alaska and the East Coast/Europe. This means that some ships won't need to pass through Panama, and thus could put the country's recent canal extension in jeopardy over the decades to come, or perhaps even sooner.

I was keen to get out of Panama, where I felt a strange coldness and disconnect with the financial calamity that had erupted immediately prior to my arrival. Why didn't I see beggars on the streets? Had they been displaced? How could there be so much apparent wealth, especially after the Panama Papers had been leaked? Was the silence a reflection of shame? The locals I saw didn't seem fazed at all. I was concerned when I began to join the dots and looked beyond Panama: globalization, financial markets, fair trade (or lack thereof), deforestation, violence, silence, and compliance. Somehow, all of these elements were connected in a complex web that I was just beginning to unravel.

Farther ahead lay the Darien Gap, a rich and extremely biodiverse break in the Panamerican Highway that crosses the North and South American continents. After making numerous enquiries, I decided not to take one of several possible routes across the relatively short stretch of water because I was concerned about the pirates that are known to operate in that area, who would have no trouble relieving me of my British passport, equipment, and cash.

With a heavy heart, therefore, I took the only other viable option, which was a one-hour flight to Bogotá, Colombia. I flew over the Darien Gap and was brought a step further toward my first major milestone: reaching the southern tip of Argentina. Even before touching down in Colombia, I had already secured several interviews, including a few related to an exciting solar-powered solution that could help reduce our dependence on oil. I was making progress on my journey and learning so much in the process. South America, its people, treasures, pains, and hope, awaited.

· 2 ·

Colombian Energy

SOLAR SHOWER

Jardin Colibri, which literally means Hummingbird Garden, is a family-run hotel about thirty miles north of Bogotá set in an intensely green and lush valley. As I sat there taking it all in, chickens and ducks roamed freely across the grounds where spinach, kale, lettuce, quinoa, corn, and a dazzling assortment of herbs were grown to provide for the family and their guests.

As I took another sip of organic coffee sweetened with locally grown honey, Mauricio Toro, the owner, told me more about his hotel: "We use new, ecologically friendly technologies . . . it's great having a solar-powered water heater . . . it's a privilege. When there is [enough] sunlight we have hot water to wash plates, to shower, which is what we need the most."[1] We went outside to look at the Kessel solar water heater, made up of a large panel covered in black material that sits at a slight incline and draws cold water from a pipe at the bottom that is connected to a three-foot-high water tank. The sun's radiation causes the water to heat up and rise through capillary action, where it eventually passes into the tank that is insulated to prevent the water from losing its heat.

Mauricio brought his electricity bills to the kitchen. We saw a modest reduction in energy use that coincided with the installation of the Kessel unit. According to him, the unit provides hot water about 85 percent of the time; however, after too many consecutive days of bad weather, he simply reverts to the electric-powered water heater, making the solution in these variable weather conditions hybrid technology.[2]

I first learned about Kessel solar water heaters when I met Samuel Neuenschwander, a charismatic Swiss engineer, in Mexico City a few months earlier. When I saw his invention, I was impressed and enamored: a simple, elegant solution to the perennial problem of burning fuel to heat water. Country manager Andres Moreno was spearheading the Kessel operation in Colombia. He was looking at ways to harness solar potential in large cities such as Bogotá (six million people), and Cali (two million)[3] in order to have the biggest impact. "There is a similarity between the requirements in the Mexican and Colombian markets, but if you look at the evolution and uptake of renewable energies in Colombia from political, social, or environmental points of view, it is behind Mexico and Brazil, but up there with Argentina," he said.[4]

Andres believed in the project to the extent that he had left his wife and children in Italy to develop the business in Colombia, where he is from originally, with a view to having them arrive the following summer so as not to disrupt the children's schooling. Our first meeting had been in Mexico City, when he had recently joined the company, so it was nice to meet him again several months later as well as a privilege to be shown a few of the installations in order to better understand the technology and business of solar-powered water heaters.

During the journey back to Bogotá from our site visit, Andres pointed out the window to the Tomine water reservoir that supplies water to the capital. Set in a beautiful valley, the artificial lake seemed to have receded significantly, perhaps half a mile or so. "We recently had the El Niño phenomenon. It's a climate anomaly which has changed the weather patterns," he explained.[5] I recalled the effects of the water shortage several thousand miles away when I was about to leave Mexico. Andres also pointed out that Colombia relies heavily on hydroelectric power,[6] which is susceptible to climate change, the effects of which I would see clearly in Zambia (chapter 13).

The Kessel system is optimized to work in tropical and subtropical regions, meaning that billions of people in the Americas, Africa, and Asia could benefit from this or similar solutions due to the position of the sun. However, the main obstacle for customers in the Global South is the capital cost of the unit, which varies across markets but tends to be about US$400–500. Therefore, the time it takes to break even—that is, for the negative cash flow to become positive—varies according to the initial price of the Kessel unit, the amount of sunlight, the cost of electricity, and the amount of water the unit can heat per day.

Although microcredits offer one path toward acquisition for lower-income users, government subsidies or energy credits would give a much-needed push to help wean the world off fossil fuels, in addition to people's own consciousness and desire to make an individual contribution to helping the climate. Compared with many products made in one country and exported en masse across the world due to low shipping costs, Andres's vision is to build a plant in Colombia and use locally sourced materials, which would not only reduce the carbon footprint of transporting materials and goods across long distances, but also create jobs locally.

The global potential for solar power is immense: according to the International Energy Agency, ninety minutes of sunlight[7] provides enough energy to cover the world's needs for a year. By tapping into this energy source, it is possible to reduce the need to burn fossil fuels, which, in turn, has several implications. First, reducing greenhouse gas emissions will help to curb global warming. Second, using renewable energy ultimately will reduce user costs over time. Third, human rights violations associated with oil extraction may lessen, an idea I will touch upon in chapter 9. Fourth, solar heaters are visible on people's rooftops, meaning that a city landscape using them changes from a polluted, car-centric one to one fit for the twenty-first century. Finally, solar water heaters have an additional feature that helps to reduce water waste: you don't need to use electricity and waste water while the tap runs, waiting for the boiler to kick in; the water already is warm, ready to be used at any moment.

Things need to change, and they can. A few weeks after leaving Colombia, pilots Bertrand Piccard and Andre Borschberg completed the first twenty-five-thousand-mile, round-the-world journey using a solar-powered aircraft called Solar Impulse. On arrival in Abu Dhabi, their final destination, Piccard declared, "The future is clean. The future is you. The future is now. Let's take it further."[8]

COUGHS, COLDS, AND COAL

A couple of local boys, who looked about nine or ten years old, carried a calf and played with it as if it were a domesticated animal, while a group of adults sat in the shade outside the local convenience store. As I looked across the worn-out soccer pitch, the empty village school in the background, I sensed something unique: a latent film of depression among the older people, their silence filling the void in the hot purgatory called home. Only a few hundred yards away, the diggers and trucks were excavating and transporting coal from the open-cut mine, throwing up dust. It left a faint whiff in the air, an acrid smell, even though according to the locals, mining had been suspended, so there shouldn't have been any activity.

The community had been waiting for a resolution nearly six years, their voices strained and nerves frayed by the collective sense of injustice about the way they felt the mining company had behaved with impunity, with only a handful of people, usually non-governmental organization (NGO) workers, truly listening to them. A 251-page report, published in English and Spanish by Pensamiento y Acción Social (Social Thinking and Action) and Arbeitsgruppe Schweiz Kolumbien (Swiss Colombian Working Group), titled *Shadow Report on the Sustainability of Glencore's Operations in Colombia*, details the negative consequences that operations in the Calenturitas coal mine have had in the adjacent town of El Hatillo:

> This report's findings include serious and repeated infractions of national environmental legislation and yet to be quantified environmental damages related to coal extraction in forest reserve areas, the unauthorized diversion of riverbeds and the exposure to air contamination levels, for populations near the mines, which exceed the limits allowed by Colombian law.[9]

Thanks to an old friend, an NGO worker, I was able to shed light on this particular mine. Sandra Froidevaux's contacts helped me learn more about the poisonous and destructive practices of the coal mining industry. They seemed to be at the center of a proverbial black hole that leeches life, fertility, and beauty from the countryside and is a key driver of carbon pollution worldwide. It seemed like the polar opposite of the

Kessel solar water heating solution I had spent time learning about. I couldn't avoid going to El Hatillo.

In the grand scheme of mining operations around the world, the plight of El Hatillo seems neither isolated nor distinct. For example, the Tar Creek zinc and lead mines in Oklahoma, in the United States, were the cause of numerous environmental and health issues spanning decades, resulting in a lawsuit over polluted water in 1934. Numerous other litigations followed, occurring well into the twenty-first century.[10]

A more recent example follows the rupture of the Fundão dam in southern Brazil in 2015, where iron-ore residue from an upstream mine was released into the Rio Doce river close to one of my stops (chapter 3), turning it brown and destroying homes and ecosystems along its way to the Atlantic Ocean.[11] The thousands of people who were affected, including fishermen, farmers, hoteliers, and anyone else whose livelihood depended on the river, no doubt, have experienced much heartache over the years that followed as they sought to rebuild their lives, hoping that the companies responsible for the disaster would remediate the plethora of issues that arose as a direct result of mining operations and the subsequent failure of a holding dam.

What seems to vary in these, and other cases, is the level of accountability and enforcement of law with regard to the responsible parties. Moreover, in countries that have higher corruption indexes and a looser implementation of environmental laws, it's perhaps easy to understand how avoidable problems can happen: Colombia ranked 96 out of 180 in Transparency International's Corruption Perceptions Index 2019[12] and attained 130th place out of 180 in the 2020 World Press Index.[13]

Meanwhile, a Global Witness report stated that 2015 was the most violent for defenders of land and the environment around the world, with twenty-six assassinations in Colombia alone.[14] The report indicated that Latin America was by far the bloodiest region on the planet with respect to defense of the land and the environment, placing it ahead of Africa and Asia.

It is patently obvious that killing environmental activists to further exploit Latin America's riches would be both unimaginable and unacceptable in countries where the rule of law is upheld and freedom of expression is embraced. A few years after my visit, a villager was murdered. Although the villagers are unclear whether this was related to the protracted dispute between El Hatillo and the mining company, I have my suspicions, given the levels of violence in Colombia and Latin America on the whole.

Colombia's coal industry has seen growth, even though fossil fuel prices (coal, oil, gas) have been relatively low in global markets in recent years.[15] Banks can underwrite loans, trade physical and derivative coal products, and, in some cases, even own shares in the mining companies, as is the case of Goldman Sachs[16] in El Hatillo. Banks, as well as energy companies (whose shareholders can also be banks), not to mention countries such as India and China who are buying this dirty hydrocarbon to fuel and accelerate their economies, all play a part in giving this poisonous industry greater longevity.

In the case of El Hatillo, the collateral effects of mining on people and the environment were stark. Several locals complained about respiratory issues, including constant colds, headaches, and itchy eyes. The river had become polluted, the air unsafe, and the land almost barren. One resident, Jaime,[17] told me, "I haven't been able to carry out my work in agriculture. It has affected my family, some have had to leave, others have lost their jobs. We used to grow everything here . . . bananas, yuca, corn, sesame, beans, sweet potato."[18]

Not only did the mining company pollute, but it acquired land. As Jaime explained, "The land wasn't ours, but we used to plough the fields. The owners, who were kind neighbors, let us grow and harvest on their land."[19] When the mining company arrived, the owners sold their land, thereby reducing agricultural land for Jaime and the other villagers. "I don't have pigs anymore because I don't have anywhere to keep them. The same is true for cows and goats."[20] Their livestock and agriculture industries disappeared overnight.

As a result of the precarious living conditions, the villagers had been negotiating a resettlement deal with the mining companies in the hope of turning their lives around; however, six years into negotiations, a mutually agreed relocation package still hadn't materialized. As another villager told me, "There is suffering in the community, there aren't any jobs, the resettlement has been delayed due to numerous complications, due to the mining company and the government, which is also to blame for this . . . there isn't enough food, even though people have food rations, they are barely surviving on that."

Tatiana Rojas, an NGO worker, invited me to a meeting she held regularly with the local committee to assist them with their resettlement negotiations. Toward the end of the session, she began to read a letter she had written to tell the committee that she was disappointed that only three people had showed up to the previous meeting on Sunday. Feeling overwhelmed, she asked a colleague to continue reading. It was embarrassing and disheartening to see a grown woman who seemed to be doing her utmost to help, sobbing with her head in her hands.

Later, I learned that Tatiana had been spending three days a week in the village while her husband worked the other three days. The single day they had off together, they spent with their child. The villagers also knew about her personal sacrifices, and that her disappointment in their collective attitude was justified. Fewer than half of those who remained at the end of the meeting sat listening with heads bowed, their faces sullen due to the reproach. After Tatiana's speech, they queued up to apologize to her.

Tatiana's stamina and commitment were unlike anything I had seen before. I was curious to find out what motivated her to make these sacrifices. "I want a different Colombia and I cannot be indifferent to what's going on at the moment . . . I believe that confronting these realities may reveal things that ensure that this isn't repeated elsewhere . . . these situations should be prevented in other parts of the world where transnationals are doing similar things."[21]

The main contention for the villagers seemed to be the resettlement issue. In August 2013, they declared, "The community will not survive if there is no relocation process. This is our third try, and we wish that it will be the last one, and one that will be dignified and successful." Their resettlement most likely would place only a modest financial burden on the mining companies' balance sheets. Yet, as I witnessed during my visit three years later, they remained deadlocked.

Interestingly, according to Vale,[22] the previous owner of the mine, they believed that the projected exhaustion date would be 2021. This meant that, all things being equal, once the coal has been exploited and mining activities have ceased, economic conditions will be even more depressed due to less employment in the area, worse pollution, and no resolution to the resettlement, essentially leaving the village high and dry.

On my last day, I spoke to Jaime and asked him what he would like to see in El Hatillo. "If I had a magic wand, I would unite the community, return the water to the dry rivers, return the birds and other animals to their habitat, for us to live in dignity with nice homes, sewage, good schools to have quality education, health care, that those mountains over there would return with vegetation, that spring would fill it with colors and a great number of small birds standing on flowers, with masses of bees flying around them."[23]

Although the narrative in the mainstream media and some politicians in the Global North imply that a smooth transition from fossil fuels to clean fuels is needed, the response is languid. It doesn't factor in the destructive nature of coal, oil, and gas, wherever they are sourced: the human environmental effects are downplayed, and the human cost is entirely overlooked.

I would leave Colombia with mixed feelings: an avant-garde eco hotel in the Colombian highlands using Swiss technology mixed with Mexican and Colombian know-how to harness the sun's immense energy was a step in the right direction compared with the bloody, dirty coal-mining practices I had seen and what was related to me. The struggle between those who want to live in a better world and those hell-bent on destroying it is a theme I would revisit during the next part of my journey through the most important fountains of life in Brazil: the Amazon Jungle and the Atlantic Forest.

Two Great Forests

COCAINE, TREES, AND A BOAT

*F*ollowing my travels in Colombia, I was left with limited options to continue my journey across the continent. In neighboring Venezuela, where shortages of food and medicine, hyperinflation, increased violence across the country (including cannibalism and decapitations) had resulted in an exodus of Venezuelans, some of whom I'd met earlier in Colombia, Panama, and Mexico, the government sealed its border with Colombia, effectively trapping its people in a huge prison.

The other overland route would have been the road south of Cali to Ecuador; however, numerous enquiries revealed that no longer was an option because the roads were closed due to strikes by farmers. As was the case when I had to fly from Panama City to Bogotá to avoid the Darien Gap (chapter 1), I reluctantly bought my second flight, this time from Bogotá to Leticia, a sketchy town on the porous tri-border with Peru and Brazil, where cocaine[1] is trafficked from Peru.[2,3]

After spending a night in Tabatinga on the Brazilian side, I changed more money (in fact, I got better than market rate for U.S. dollars to Brazilian real from currency dealers on the street) and got my passport stamped at the police station at the other end of town. Then I bought a hammock and basic supplies for a three-day, seven-hundred-mile boat trip downstream to Manaus, the capital of Amazonas state—a place I had long wanted to visit for the strange fact that it's the most populous city in the Amazon rain forest. I made my way to the dock, where a queue had already formed two hours before the boat's departure, and stood in line, thinking about what lay ahead. A strange mix of excitement, curiosity, and apprehension filled my mind.

This was by far the cheapest mode of transportation that I would take on the whole trip across Latin America: for the equivalent of about US$42, I would have transportation, food, and lodging for three days and nights aboard the *Manoel Monteiro*. It is the only form of public transportation for isolated communities, cargo, and a few tourists along the Amazon River, with one of the most impressive and important natural habitats in the world lining its banks: the Amazon rain forest.

There were, however, a couple of flies in the ointment. After doing online searches prior to the trip and speaking to a few people onboard, it became apparent that a possible sinister scam in operation could have turned my trip into the worst possible nightmare: drugs being planted in my luggage to get *coima* (a fine) out of me, or worse, which might involve getting my goods confiscated. Ruminating on this took me back to Nicaragua (chapter 1) and the close call with the motorcycle thief that had so shaken me.

Another issue: I was heading into a region[4] where the Zika virus, known to be transmitted by mosquitoes and to cause a range of symptoms (from a mild rash or pain to paralysis and birth defects), was active. As I had already lived through the H1N1 swine flu emergency in Mexico City in 2009 that had caused death and suffering, I decided to take a calculated risk. I simply applied insect repellent, continuing to use my malaria prophylaxis because I felt that malaria was a bigger risk anyway, and crossed my fingers that I wouldn't be bitten by an infected Zika mosquito. A few years later, the coronavirus would emerge and have a rather big impact on my journey as I drew toward the finishing line (chapter 16).

To compound my apprehension, when I boarded the boat, a Swiss and Italian man who had just eaten magic mushrooms and were tripping at 10 a.m. tried to socialize with me. At face value, this made sense—they were on a psychedelic trip in one of the most beautiful places on the planet and probably feeling quite euphoric. My problem with them, apart from their strange behavior, was that they might plant drugs on me. I mean, if they were on drugs, then such an idea wasn't far-fetched. I politely excused myself, preferring to pitch my hammock in a more shaded part of the boat.

Over the coming hours and days, my excessive concern slowly dissipated into calm, and I befriended people on the boat. In particular, it became obvious that João, a father of three who was sleeping with his family right next to me, was a good man and that I could trust him to look after my belongings when I had to go to the bathroom. However, at times my calm was short-lived, such as when I experienced the second major shakedown on the boat. Both federal police and soldiers came aboard as though the boat was a film set and the most senior guy was Bruce Willis who'd had a bad day. It made me wonder whether they were searching for drugs, looking for a bribe, or, worst of all, performing a theatrical piece to give the illusion of a war on drugs. In any case, surely the drug cartels already had this covered.

Spending three days and nights watching the same landscape, trees and water interspersed with the occasional serene village where a few families were farming and fishing, was transfixing. The intermittent intense showers, sunrises, and sunsets are what dreams are made of; such is the beauty I was beholding. The river swells and contracts on a yearly basis, much like a lung does with each breath. My trip occurred a few weeks after the river had started to drop, and I enjoyed views of some smaller islands that could be seen jutting up, which a local on board pointed out to me.

Today, alongside the environmental dangers of deforestation, the Amazon, and Brazil in general, is the world's backyard for meat production. The official data I reviewed did not specify whether the meat came from the Amazon; however, millions of tons of Brazilian meat are exported across the world to places as far flung as Russia and China.[5] In doing so, as was the case with the cows I saw grazing outside the Naha jungle in Mexico, massive amounts of CO_2 and methane are being released into the atmosphere in conjunction with the CO_2 release of the felled trees and the additional carbon footprint resulting from the transportation, which, for the aforementioned countries, needs to reach the other side of the planet.

Considering that, my plate on board the boat, with the exception of some locally caught fish, the food was vegetarian—beans, rice, salads, potatoes, and fresh fruit. Several studies have pointed to a logical argument that becoming vegetarian could cut food emissions significantly. But even this is contentious, as I would learn later, because key vegan foods such as soy and avocado involve deforestation, not to mention the human suffering caused by land grabbing, as I would witness further along my voyage across South America (chapter 5). In any case, reducing meat[6] and dairy consumption, it can be argued, is a fairly easy, quick win, because cattle farming is a key driver of deforestation in the Amazon[7] and carbon emissions.

On the final morning, I sat on the top deck to write my notes for the day, and smelled the strong fragrance of hashish that some Colombian passengers had brought with them. Smoking cannabis in public has become normalized in some parts of the world, from hippie beaches on the Pacific Coast of Mexico to laid-back neighborhoods across the Global North. Cannabis, which is also intrinsically linked to organized crime, has become a normal, almost quotidian part of today's landscape, so in some ways I wasn't surprised.

I reverted back to my notes to jot down all of the small details I could see as the river began to narrow and the currents gave the boat extra pace. Houses were no longer sporadic, but lined up almost continually along the banks of the river, raised on stilts, where bananas, corn, and some livestock could be seen in larger numbers. I could see many farmers with translucent plastic backpacks filled presumably with an agrochemical such as an insecticide or fungicide, poisons with which I would come face-to-face in Africa (chapter 12).

Finally, Manaus, the largest city in the Amazon and home to two million people,[8] came into view. The dock was filled with dozens of taxi drivers shouting out different destinations. I was herded into the back of a two-seat pickup truck with a Colombian preacher who, in rather poor Portuguese, told us that we should find the path to God and that the biggest threat today was fundamentalist Islam. The taxi driver nodded politely, and I just looked out of the window at the military airport, crowded skyline, and penetration of technology in the form of mobile phone reception, Google street view (thanks to Google cars), and Uber taxis.

The traffic jams, the noise, the banks, the supermarkets—all seemed out of place in the middle of a tropical rain forest, not to mention the opera house in the city center. It seemed an infestation of humans. I could see people everywhere. It was surreal to believe that I was in the middle of the Amazon after having had a vantage point from which to travel through the Earth's lungs at a snail's pace and observe the beautiful forest that is home to a small number of people, among

them some with whom I had shared the boat. But all of the previous serenity and beauty was at risk. What would the future bring? How could
deforestation be reversed? How could I find hope when all of the data I
had reviewed about the future of trees pointed to something as sinister
as the total collapse of the climate and, by consequence, a significant
number of other species including *Homo sapiens*?

SALT OF THE EARTH

We might feel optimistic when we hear about bioethanol fuel. It's
touted as being a green renewable energy source. A simple process
converts sugar to ethanol (alcohol), which is later added to fuel. The
percentage of ethanol content in the fuel can mean significant reductions in CO_2 emissions at the tailpipe of suitable cars. This, in principle,
sounds good because, theoretically, more sugarcane can be harvested
from the same plot of land, meaning that even more CO_2 emissions can
be reduced over time.

A familiar sight throughout Brazil, both on this and previous trips,
were bioethanol pumps at gas stations, alongside normal gasoline and
diesel. While bioethanol sounds like a good alternative to traditional
fuels, large swaths of forest are being felled to facilitate it. So, as in the
case of cattle farming, the first consequence is that CO_2 is released from
trees when they are cut down; then ecosystems disappear, and soil is
eroded. The cruel irony is that, after all that destruction, the myth that
biofuels are good because they're supposedly "green," which car makers
are so desperate to fool their customers into thinking, was debunked by
a group of scientists who came to the following conclusion after com-
paring bioethanol production in the United States and Brazil:

> The use of ethanol as a substitute for gasoline proved to be neither
> a sustainable nor an environmentally friendly option, considering
> ecological footprint values, and both net energy and CO_2 offset
> considerations seemed relatively unimportant compared to the eco-
> logical footprint.[9]

Much of Brazil's ethanol production occurs in the south of the
country,[10] which overlaps with the Atlantic forest. Deforestation in this
lesser-known forest, which covers a vast area that extends as far south

and west as Paraguay and Argentina,[11] and east toward the coast to places such as Rio de Janeiro,[12] threatens trees, plants, and animals that cannot be found anywhere else in the world. This, together with continued urban expansion, puts the remaining 7 percent of the Brazilian part of the Atlantic forest under extraordinary pressure.[13]

Set in a valley within what remains of the Atlantic forest more than fifteen hundred miles southeast of Manaus, I finally arrived at Instituto Terra (Earth Institute) located just outside the provincial town of Aimores. The grounds surrounding the institute were featured in *Salt of the Earth*, a documentary film[14] about the life and work of Sebastião Salgado, a former economist and development banker who in his thirties discovered photography. Today, he is considered one of the world's most prolific photographers; his body of work encompasses subjects as diverse as coffee farming and the Rwandan genocide. After Salgado was bequeathed a dried-up farm from his parents, he embarked on a project with his wife, Lélia Wanick Salgado, to bring his part of the Atlantic forest back to life.

The Oscar-nominated documentary, codirected by Wim Wenders and the protagonist's son, Juliano Ribeiro Salgado, is a testimony to renaissance, to rebirth, to not giving up on something that apparently no longer has any use, to look for a solution even though the path might be long and uncertain. A forest, apparently dead after a drought, was brought back to life over a twenty-year period through a program of extensive tree planting.

The executive director of Instituto Terra, Isabella Salton, a former sales director in a transnational firm, gave me more background about the institute's trees' ability to support life and sequester 100,000 tons of CO_2 a year: "We have planted more than two million native Atlantic seedlings. We want to restore the natural diversity, so we need all specimens here." Scientists collect two hundred different types of seeds from within a 120-mile radius of the institute, analyze them in the onsite laboratory, and later develop them into seedlings.

In addition to developing their piece of the Atlantic forest, which provides a home to more than 33 different mammals, 107 birds, and 16 reptiles, Isabella told me about another initiative that will help safeguard the future of the Rio Doce basin, which recently was hit by a catastrophic mining disaster upstream, releasing iron ore residues from the Benito Rodrigues dam that broke the previous year. "We created a special program called *olhos de água* (water springs); we take care of the springs in order to put more water in the main river. Our objective is to restore the water in all of Rio Doce basin."

The program began in 2010, five years before the mining disaster, making it doubly important today. "We learned in the last five years. We did around one thousand [springs]. We want to do five thousand in the next five years." The objective is to have three hundred thousand springs within thirty years, which will help restore this rich, diverse ecosystem.

My last stop on the forest tour was the nursery, where staff members were working on the massive allotment, about the size of a soccer field: thousands of seedlings, arranged in neat rows, had name tags with their dates of birth on them, resembling a neonatal ward in a hospital. After taking photos of the endemic Garapa (*Apuleia leiocarpa*) and Curindiba (*Trema micrantha*) tree seedlings of varying heights, colors, and ages, I reflected on what I was observing: the Naha jungle in Mexico and the Amazon rain forest were at great risk due to the radical change in their ecosystems. Although this troubled me, being surrounded by life inside the Atlantic forest on the absolute brink of extinction gave me some hope, but later it would create a conflict in my mind.

The fruit of the institute's hard work showed me what is possible when the human will to identify and treat a problem exists. If only more people could see this. Part of today's problem is a collective paralysis because people are overwhelmed by the constant barrage of negative news in the press, social media, and elsewhere, making it easier and more comfortable to switch off rather than enquire and discover the solutions, and that we, as individuals, can be a part of them.

The macro solution will need to be a collective one. It cannot be left in the hands of a few concerned individuals. For example, although Norway recently banned deforestation, thus representing an important sovereign state putting theory into practice, a few years later it decided to continue drilling for oil, demonstrating climate hypocrisy, which I will expound upon in later chapters. If more states, corporations, and individuals help to reverse the current destruction, then it might be possible to imagine a better future, but we need congruence and consistency.

As I stood on the hill among the young Garapa trees, I inhaled the damp, earthy smell from the ground and looked down at the beautiful reception center that Salgado and his wife, Wanick, had converted from a barn. The pictures that I had seen of the site decades ago showed a barren, almost extinct wilderness; yet, there I was, a witness to nature's miraculous strength. But something else I couldn't see was happening beneath my feet.

Trees are known to communicate with one another through a complex network of mycorrhizal fungi, the Wood Wide Web. They can transport water and nutrients. Can we humans, custodians of forests, learn from their altruism, and find ways to communicate better and share resources with one another, and through this, live on a more balanced, beautiful, and harmonious planet?

Every time we click "save" on an e-mail, document, or spreadsheet, we are asking the computer to save binary information (ones and zeros) to memory. That memory, for all intents and purposes, will persist forever. As I stood surrounded by the trees, like a baby in its crib surrounded by a blanket, I wondered how many memories the fungi had. Did they know that they had died and come back to life? Were they aware that the team at the institute had searched for their siblings in a 120-mile radius to promote diversity? I didn't need to ask if they had had enough to drink or needed any nutrients. The easiest measure was to hold their trunks and gently press their deep green leaves. They were in good hands.

Being immersed in nature, with its rich palette of colors, sounds, and textures set on a canvas of exquisite beauty, would be a useful respite from months of intense traveling, during which I had to be on a state of high alert most of the time. The next stops would be two large cities in South America that face other challenges, where other ingenious solutions have been devised, followed by an entirely different rural landscape in Argentina. I would leave the verdant and tranquil forest with a warm heart and enriched soul.

Ingenious Shantytowns

RETURN TO THE SHANTYTOWN OF PEACE (FAVELA DA PAZ)

$\mathscr{A}$ long, blue flame came to life when Fabio Miranda used his cigarette lighter to show off the power of his new biodigester system, a large, gray plastic cube about four feet high. Fabio showed me the inside of the cube and explained that food waste from seven homes is stored inside the device and processed by anaerobic bacteria, which release methane gas that is then provided to three homes. This has meant that Fabio hasn't needed to buy gas for approximately a year and a half.[1]

I had met Fabio three years earlier at his home in Jardim Ângela, a densely populated favela (shantytown) on the outskirts of São Paulo, one of the world's most populous cities, while doing fieldwork for my first book, which looked at how music could be used to overcome societal problems. Fabio welcomed me with open arms on that first visit. He not only showed me his studio, but also gave me a tour of his home and the whole favela. We kept in touch via Facebook, and I learned more about his other passions, electronics and permaculture, so it seemed entirely natural for me to return. This time, however, we wouldn't be talking about music but the environment and what he was doing to improve the situation for his family and community.

The biodigester in Fabio's workshop has the ability to reduce his family's carbon footprint. It was central to a theme that I had been contemplating since I saw the destruction of trees in the Lacandon jungle in Mexico a few months earlier (chapter 1), together with the challenges faced in the Amazon and Atlantic forests that I had visited just a couple of weeks before my arrival in São Paulo. The biodigester was the antithesis of the cycle of consumerism that doesn't attribute value to waste. Fabio has the capacity to find the intrinsic value of waste and transform it with his ingenuity.

Another important function the biodigester performs is the creation of biofertilizer, which Fabio explained is used in the organic hydroponic garden where he grows lettuce and parsley, staples in his family's largely vegetarian diet.[2] This is striking for a number of reasons. First, by using hydroponic technology, he can grow vegetables using less water. Second, by using fertilizer from his machine, he doesn't need to buy fertilizer. Last, having rows of hydroponic vegetables stacked vertically upon each other is a more efficient use of space compared with the traditional soil method; he can grow more vegetables in the same area—a useful consideration given the extremely limited space. That limitation came into even sharper relief when I took in the view from the garden: we were right on the edge of the favela and, as far as the eye could see, small houses were built precariously, one on top of the other, with large red bricks, cement, and metal roofs, in what could only be described as overcrowded, cramped conditions.

Numerous other innovations in Fabio's home continued the hybrid theme. Through an intricate network of metal gutters and plastic pipes, rainwater is collected in two tanks (one 100 gallons, the other 265 gallons). The larger one is connected to the home's main water supply and, using a fill valve and float ball, always ensures that the larger tank has at least one hundred gallons. By having two tanks, Fabio and his family can profit from excessive rainfall and store it for later use, such as washing dishes and clothes, as well as watering the plants dotting the house, making his home feel like an oasis in a concrete desert.[3]

After I told Fabio about the Kessel solar water heater system that I had encountered in Colombia (detailed in chapter 2), he pointed above us: "We've got a similar thing on the roof."[4] We climbed up a wooden ladder and stood on the roof, where a two-hundred-foot black rubber hose connected to the smaller water tank. Due to its length, color, and the material it was made of, the hose was hot to the touch and transferred heat to the water that slowly passed through it. In other words, a Swiss inventor living in Mexico and a Brazilian inventor living in Brazil had found similar solutions to the same problem.

"There's something that I want to show you," he said with a sneaky smile, so we climbed back down, and he showed off the rest of his workshop, demonstrating a tiny piston engine about an inch long he had developed that runs on biogas. As a proof of concept, it worked well: it produced enough electricity to light an LED bulb. I asked Fabio why he was doing this; he replied, "It's possible to produce this form of energy . . . I'm a musician . . . I can add value to the lives of others, for the whole world. Everyone has this energy inside of them, it's just a question of finding it."[5]

Indeed, his capacity to transform inanimate sheet music into waves of sound that enrich the lives of others is similar to what I saw when he lit a bulb using electricity produced from decomposing waste. His enthusiasm seemed unabated as he showed me a selection of solar panels, ranging from about eight inches by six inches to the largest one, at about thirty-five inches by twenty-five inches, which produces approximately fifty watts of power to drive the water pump in the house. Excess electricity is stored in two batteries.

The following day Fabio took me on a tour of the favela. As we walked past the local primary school, Fabio told me that his NGO, Favela da Paz (Shantytown of Peace), is working on a nursery in partnership with the health ministry to help encourage young people not only to learn about gardening, but to get into the habit of eating healthily.[6] Developing these skills will be useful tools to tackle obesity—a major problem in Brazil, where 20.1 percent of the population falls into this category, according to the World Health Organization (WHO).[7]

As we continued walking through the narrow, steep, and rather haphazard streets, I couldn't help but notice that things seemed to look better than when I first visited the neighborhood three years earlier. Fabio explained why that could be the case: "Businesses are growing here. Before, if I needed to buy a guitar string for my bass guitar, I would have to go to the center of São Paulo, but now I can buy it right here in the favela."[8] It's as if the favela itself is developing and maturing: I saw a brand-new car outside one of his neighbor's homes, which Fabio attributes to hard work: "We work many hours and have numerous jobs."[9]

The cliché image we see of Brazilian favelas when we see them on television is that, in spite of their hardships, music, dance, food, drink, and lots of colorful buildings help them overcome their reality, and that they just seek entertainment and fun to find their way out of trouble.

But this couldn't be further from the truth. Although people I spoke to in the favela believe that violence has decreased, some still deal and consume hard drugs such as cocaine. Robberies, petty crime, poverty, unemployment, and other uncertainties are part of living in the favela, but in spite of this, life goes on because the human will to survive drives them forward.

In light of this, I wondered whether Fabio's cheeriness may be seen as a positive coping mechanism to help him deal with the anxiety that comes with the societal and economic hardships entrenched in his neighborhood. By connecting good intentions with hard work, Fabio and the other members of the Favela da Paz collective whom I had met on this and my previous visits had brought about positive changes to their own lives and the wider community through a range of activities, from music and video production to permaculture. They have grown from an organization to an institute that benefits from government support, as well as recognition both from within the favela and beyond.

We have a lot to learn from each other. Not only are Fabio's water heater, biodigester, solar power, and other inventions desperately needed in other parts of São Paulo, but also in Mexico City, Nairobi, Soweto, and an endless list of other places. His sense of duty and commitment to understanding the climate crisis and looking for ways to overcome it were shoots of hope that had sprouted because he recognized the enormity, gravity, and interconnectedness of the issues at hand, even if the underlying complexity is too great to be understood by a single person. He found his own solutions, essentially turning potential energy into kinetic energy, reflecting his love and commitment to his family, community, and the world.

Fabio often used the word *irmão* (brother) when he spoke with me. Naturally, I reciprocated. He introduced me to his family and opened his home, heart, wisdom, and light to me. The wonderful moments we shared will stay with me forever. If Salgado's forest I visited in the Amazon (chapter 3) was a grain within the salt of the Earth, then surely Fabio is part of that life force, another element in a rich tapestry that makes the world a better place. For the communion of love, fraternity, and devotion is the only thing that can possibly counteract the malevolent forces that oppress, hate, and destroy what is so dear and precious to all of us: life itself and the bright colors that make it so beautiful.

RECYCLED ORCHESTRA

"She practices from Monday to Monday," said one of the mothers, Viviana, bearing witness to her daughter's joy and enthusiasm when playing the violin. "The orchestra represents the best, the most beautiful things. It changes the kids' lives in my view, because they're not thinking about other things. They're thinking about music. This is a neighborhood that has problems, violence, drug addiction."

The Recycled Instruments Orchestra of Cateura (*Orquesta de Instrumentos Reciclados de Cateura*) has used materials from the nearby landfill to make its own instruments in the onsite workshop. It was catapulted to fame through a variety of means, including being the opening band for Metallica during their South American dates in their *Metallica By Request* tour in 2014; the star of a feature documentary called *Landfill Harmonic*, which has enjoyed widespread success at film festivals around the world; a performance for Pope Francis during his visit to Paraguay in 2015, where Viviana's daughter and other members of the orchestra played for him; and the founder, Favio Chavez, having given two TED talks in Europe.

During the afternoon I spent at the school, I heard students' efforts manifest themselves as a constant drone of musical phrases being repeated over and over again, with violins in one room, cellos in another, clarinets in another, and so on. The uncoordinated clash of sounds created an acoustic melée requiring even more discipline for each musician to focus on what he was supposed to do: strive for individual and collective perfection of the Bach violin concerto that the orchestra would perform in a few days' time with American conductor Dr. Glenn Block, director of orchestras and opera and professor of conducting at Illinois State University.

Meanwhile, men were working on the extension to the school in order to increase capacity to meet high demand both within the community and farther afield. Marcio Weber, twenty-four, who had been in the orchestra for ten years, still visited the school regularly and believed that the orchestra changed his life: "I started with normal instruments, but it was a recycled instrument that allowed me to travel the world."[10] The list of countries and regions he visited while touring with the orchestra is impressive, including Abu Dhabi, Canada, Central and South America, and "around eight or nine countries in Europe."[11]

Marcio kindly showed me around the school, which has about ten classrooms. We stopped in the storeroom to look at a mix of recycled and standard instruments, then visited the workshop. The creativity I saw there was breathtaking: a banjo whose body was made from a saucepan and strap from an inner tube; a double bass made from an old oil drum; numerous violins that had forks previously used for making gnocchi as the tail pieces to hold the strings. I asked Marcio if he could play something on one of the recycled instruments. The Bach sonata he performed on the recycled violin with a metal body and wooden neck sounded slightly muted, but not metallic. He had transformed waste into music.

The sounds of baroque music on a desperately humid day next to a landfill in Paraguay should seem neither far-fetched nor contrived. Their repertory, just as the means by which they make instruments, is reimagined, and it was contagious. The school was abuzz, parents, students, and I all aware that the musicians were doing much more than converting trash into something useful: they were changing their environment, harvesting hope and opportunity from things previously thought of as useless, surplus. At the end of the rehearsal, a volunteer from a nearby school came to pick up three guitars and three violins to complement existing instruments. As one of the mothers told me before I left, "it's the best thing to ever happen to our community."

I visited Favio Chavez's home. He founded the orchestra and explained the inspiration behind it: "The project began ten years ago within the community that lives around the Cateura rubbish dump. It is a poor community with children who have limited resources and opportunities. We began at a point where they live."[12] When he thought about combining a social project with his love for music, he drew upon his previous knowledge as an environmental technician in the landfill and combined these elements to create the orchestra. The extreme living conditions within Cateura made him think creatively because no resources were available, including musical instruments, with which to teach the children music.

As I moved the conversation toward the authenticity of the orchestra, Favio reflected: "It is authentic because it's related to where they come from, and it doesn't negate the conditions. It denounces the conditions that these youngsters live in. But the solution isn't to evade or escape from the community. The solution is to improve the community from within."[13]

Reusing trash to create instruments with which to perform classical repertory was by far the most original thing I had seen during the trip to that point. Favio believed that originality in this context needn't refer to the fact that they were reusing materials; rather, he saw the novelty in creating something brand-new starting at the physical root of the community: the landfill. They use the same waste material that, in many cases, is used to build their homes.[14]

Favio went on to explain that a new instrument, which costs US\$200 or US\$300, is more expensive than their home, so he believed it was inappropriate to make a child responsible for an instrument worth more than their dwelling. Notwithstanding the school's constraints, "We still try to ensure that they receive the best musical education possible within this context," he said.[15] Favio wishes to equip his students so that they not only become the best musicians that they can be, but that they also acquire a wider skill set, including fluency in English.

"This has grown gradually until the point that I dedicate myself full time to this. Today I dedicate all of my time to this," Favio asserted. This includes constant touring as well as the days spent running the orchestra—days that could have been spent with his own family, but, in spite of this, he feels "that in many cases . . . these children are part of my family as well."[16]

After completing the interviews and writing up my notes, I felt a small sense of relief. I genuinely felt that what I had seen was remarkable, especially after the tensions in Central America and things I had seen in northern Colombia and the Brazilian Amazon: I had the privilege of seeing examples of hope. We desperately need to celebrate and develop projects such as this one in order to inspire and involve more young people, to help them envision and shape a cleaner, brighter, and more prosperous future.

The ingenuity, both at the recycled orchestra in Paraguay, and earlier, at my dear friend Fabio's home in Brazil, reflects the human being's capacity to look for real, practical, clever, innovative, and immediate solutions even when they might seem improbable. Rather than wait for help to come, which often doesn't arrive, these bright hearts and minds have decided to empower themselves by shaping their own destiny. This, together with the winter sun that shone on me, raised my spirits, and gave me a much-needed energy boost for what would come after crossing the border into Argentina.

· 5 ·

The Open Veins of Latin America

LAND GRAB

"*Why* don't you negotiate with this guy?" said the portly man with white hair. He seemed charming and tried to befriend his audience, which had begun to increase in number, curious as to what this outsider had to say and, more important, what his intentions might be.

He continued, "How many hectares do you have?"

"Four thousand hectares," shouted one of the men who was wearing a gray hooded top and baseball cap. The thick mist that carried his response disappeared rapidly.

"Maybe you could get four or five million pesos (US$260,000 to US$330,000) for it."

"Man, if we wanted the money, we would have left a long time ago," was the agitated response of another villager. They began to walk away, tired of their unwelcome guest who, it seemed, was a trojan horse: "The guy's not from these parts; he's from Santa Fe [a city almost four hundred miles away]."

My initial query upon arrival in Argentina had been related to how peasant farming could be a viable alternative to the proliferation of genetically modified organisms (GMO) and industrial agriculture. Rather than learning about techniques and traditions passed down through generations, I discovered oppression, fear, and destruction. I had joined these farmers and their families at a makeshift tent in the tiny village of Bajo Hondo, which looked similar to images of refugee camps in present-day Europe. It was the middle of winter in northern Argentina. The featureless plains invited cold winds to blast through the improvised dwellings. The villagers' homes had been burned to the ground less than ten days earlier, their freshwater well poisoned, and the grain stores for their livestock demolished. Why would dozens of men from another province feel a compelling need to burn down their homes and destroy their livelihoods? One villager seemed to think that a powerful businessman wanted the land, no matter the implications for local life.

While the men checked their rifles, they passed around a few one-liter cartons of red wine, guzzling it to anesthetize against the cold. I moved back inside to check on the women and children huddled around a small fire as they prepared a hot mate herbal drink, which is common across Argentina. The children didn't seem too concerned despite the magnitude of the situation. Everyone seemed to exude stoicism and defiance, regardless of age or gender.

I wondered what I could do to help. I'm not a fighter. What would I do if the thugs returned? I remained silent until 4 or 5 a.m., when fatigue finally overcame my racing mind, and I managed a few hours of shut-eye. I was the first person to awaken. It was 6 a.m. Dawning gave to the dark blue sky above me a golden and orange tinge, the stars continued to fade out, one by one, and I smelled strong smoke from the campfire that had kept me warm at night. Pigs, hens, and a dog were roaming around, also wondering what was going on.

"This is where I used to live," said Carlos,[1] as he stood on a mound of ash mixed with mud. "Two coach loads of men arrived, they started throwing my belongings out of my house, and then they set it alight. It was made of wood . . . I don't have much of an education. I don't understand why they want to take our land when they have so much." We walked away from the house toward the well that Carlos had wanted to show me since the previous evening. I peered into the well to see green stains on the red brick walls. According to Carlos, they were caused by vandals pouring poison into the water to kill the livestock: "They dumped bricks in there to finish us off."

Sometimes, when disaster strikes, an unexpected show of solidarity emerges. I had arrived the night before with Chloe,[2] a middle-aged woman, and Pablo,[3] a young man, who had come from another village to bring much-needed drinking water, sugar, mate tea, coffee, oil, pasta, and flour. I met them through a mutual acquaintance with whom I had spoken a few days earlier in Quimili, the nearest larger town, where I had arrived by bus from Paraguay. My contribution was to pay for the diesel fuel to get us there and back.

After taking photos of what remained of the village, essentially several burned-out homes adjacent to the farm, I took a long walk with Chloe. She seemed to bring calm and composure to everyone, including me. A woman with warm brown eyes, she seemed to have a wealth of life experience and was constantly preparing mate tea, cooking food, and generally supporting everyone with her inimitable charm. Reflecting on what had happened to the village, she commented, "Solidarity is related to something that lives inside of us, from within. I believe that each and every person should feel a strong desire to accompany another person at times like these."[4]

Chloe also believed that solidarity shouldn't be confined to what happened in Bajo Hondo. Rather, it should also apply to gender, racial, or any other type of discrimination where vulnerable people need support: "We didn't have the opportunity to go to school in order to be able to understand what our rights are."[5] This is why she created workshops to help locals understand their legal rights and how they can be used to protect themselves, as she went on to explain: "There are lots of nice laws, but it's very sad that they are not implemented."[6] Although Chloe felt discriminated against by those who have power and money, she also recognized the power of mobilizing her peers to overcome this existential threat, which is the main reason why she came to Bajo Hondo, despite doing so at great personal risk.

Regrettably, the recent landgrab wasn't Chloe's first experience of this. She has supported other villages that have been destroyed by similar landgrabbing attempts, in which people have been forced out of their homes so that the land could be used for financial gain. She reflected vehemently on these experiences: "I still can't believe that there are human beings that don't have an ounce of sense, a morsel of soul, and can burn a farm down, that leave families outside [in this cold weather], children without a roof over their heads, that destroy their food production. These things make me angry."[7]

The cycle of intimidation, attack, and eviction is, unfortunately, familiar to Chloe, who acts as a representative for many people's sentiments in Bajo Hondo. She went on: "The justice system is unjust . . . We go to the police station, and they don't want to review our case. We try to file a complaint, and they don't give us the paperwork . . .

sometimes we end up in jail because we're the ones denouncing a crime. Why? Because we're farmers and we're uneducated."[8]

Given the high levels of corruption and social inequality, bribery is a staple in the country and on the continent as a whole. Add violence to the mix, which historically in Argentina has included the disappearance of an estimated thirty thousand people during the Dirty War (*guerra sucia*) between 1976 and 1983.[9] Another explanation for the reluctance of the police to assist Chloe may be collusion with the perpetrators or fear of reprisals, which, based on the scale of the destruction I saw, would not be difficult given the power and reach of these criminal forces.

The environmental consequences of this pattern of savage behavior were all too clear for her: "The fumigations are killing our plants and animals. They use planes to do this."[10] Chloe believed that the landgrabbers' agenda is simple: eliminate traditional farming, then replace it with GMO/high-intensity monocrops that work optimally with chemicals.[11] Naturally, all of this favors foreign currency receipts and stimulates the economy through the sale and use of industrial pesticides, herbicides, and fungicides to the detriment of peasant farmers and the environment. In spite of all of this, Chloe has stuck to her guns. Being committed to this cause has given her more energy to move forward so that one day the destructive cycle of intimidation, abuse, and violence will end.

It was hard for me to fully digest and understand what had happened in Bajo Hondo. I arrived a few days after the attack, and the smell of ash that blew through the camp hung in my nose for several days. Do all attempted landgrabs look like that? What would I have done if the attackers had returned and fired at us? During the ride back to Quimili, we drove past a lot of soybean fields. They all looked green and uniform. No way were those farms being run by peasant farmers—they were organized, big-business farms with plenty of money.

By the time I arrived at the main bus station in Quimili, my nerves were frayed. I'd had enough. I sat outside, the winter sun warming my skin but not my soul. I felt crushed. I watched two stray dogs run around a dirt patch in front of me and seriously wondered, for the first time during my trip, if I should call it a day. I mean, I'd had more successes than failures in my life, and nothing is wrong with quitting. Would a greater awareness of the challenges people such as the villagers of Bajo Hondo faced really change their reality? I drew some hope from the thought that, maybe, by documenting their resistance, this book could reflect back to them their own dignity and inspire others to resist the dark, malevolent, and pernicious forces that threaten humankind's future.

Then I recalled the day I had arrived in Quimili a week earlier, on July 9, Argentina's Independence Day. In the main square near the bus station, I drank Quilmes beer and ate freshly made hot dogs with locals while we watched fireworks, which drowned out the drums and guitars that played in the background, celebrating two hundred years of independence from Spain. I felt excited and motivated at the thought that, surely after two centuries, the people of this land would be freed from violence and oppression. Simultaneously, the farmers in Bajo Hondo were putting out fires and defending their land.

After the joys of having seen the work of the recycled orchestra in Asunción, reuniting with my friends in the favela in São Paulo, and exploring Salgado's forest in Brazil, I rationalized this bleak experience as inevitable. Even though I had seen injustices in different manifestations, constant poverty, and had had to protect myself from the violence that was endemic during my journey to that point, my innocent question regarding soybean production, which had stemmed from the simplistic link between Argentina being an exporter to China of a staple ingredient, had unearthed a disastrous and avoidable mess. The vindictive and callous actions whose aftermath I witnessed made my visit to Bajo Hondo particularly indelible.

As was the case with the farmers in the border town in southern Mexico who protested the water shortages and the calamity that Colombian farmers living next to the coal mine faced, this latest tragedy in Bajo Hondo, too, basically went unreported, apart from a few left-wing articles,[12] which ostensibly are generally read only by their affiliates, meaning that most people are unaware that these terrible things ever happened.

It is impossible to accept that good-natured, hardworking people—or anyone for that matter—should be subjected to this. The rewards of taking over someone else's land are all too clear: global demand for soybeans is strong.[13] They are used for livestock feed, sales of which provide exporters the chance to get hard currency receipts in a country whose currency, the Argentine peso, has been devalued several times since it lost parity with the dollar in 2002. Replacing peasant farming with industrial farming increases the use of GMO, pesticides, and lots of water, and it results in habitat loss and innumerable collateral effects when birds, insects, and other wildlife lose their homes and sources of food and water, concluding their demise. In simple terms, this is precisely what is referred to as the sixth extinction.

Mauricio Macri, then president of Argentina (2015–2019), listed Ayn Rand's *Fountainhead* as one of his favorite books. The central thesis of the book is objectivism, advocating the achievement of one's objectives regardless of the cost. The book's hero, Howard Roark, an uncompromising architect who dropped out of university, destroys a building that he had designed because he is unhappy that his blueprint had been modified, reasoning thus: "The first right on earth is the right of the ego. Man's first duty is to himself. His moral law is never to place his prime goal within the persons of others. His moral obligation is to do what he wishes, provided his wish does not depend primarily upon other men."[14] In line with this, Macri's objective seemed very much to make as much money as possible through continual expansion of the Argentine market without regard for the people who would suffer.

The people of Bajo Hondo represent a small number within a large group of people at the center of the impact caused by the violent and ruthless sides of capitalism and global markets. Further, if soybeans end up becoming a meal for livestock thousands of miles away, this introduces innumerable other stresses on the world's already fragile environment through international shipping and deforestation to make way for the livestock that will consume the soybeans. This, in turn, invariably means further habitat loss and pollution, together with the possible displacement of peoples. It highlights the interconnectedness of deforestation, industrial farming, agrochemicals, international shipping, global markets, and human rights abuses.

Successful landgrabs mean a good paycheck for the thugs, increased revenue for the agrochemical industry, low-priced consumer products, either directly in the form of a vegan soy product, a metabolized version such as beef, or a biofuel. In the process, the wisdom of peasant farmers becomes diluted to the point of disappearance, while the sounds, colors, smells, and richness of the original ecosystems silently vanish.

What I saw in Bajo Hondo was protracted pain and suffering at the hands of an invisible force, which, it seems, is embodied by a wealthy Argentine businessman wreaking havoc in his country. How patriotic. But he is helped and supported by key apparatus in the form of the government, banks, agroindustry, and others, who are all too keen to support the expansionist agenda that is part of Latin America's modern history. Perhaps the bulk of his wealth is in Panama, Switzerland, or some other foreign land, because I very much doubt that he would keep much of it in Argentine pesos.

But there is only one destiny; the interdependencies across our world, as I was learning on my journey, can lead to only one of two possible conclusions: life or death. The valor and integrity of the men, women, and children in Bajo Hondo was an important demonstration of the strength of the human spirit. They defended against a landgrab and still were capable of maintaining the human and universal values of respect, solidarity, and compassion, which were more alive in that village than possibly anywhere I had visited to that point.

GALEANO'S OPEN VEINS

Parque Rivadavia, a small park in a middle-class neighborhood, is testament to the voracious reading habits of Porteños, the inhabitants of Buenos Aires, where every day tens of thousands of books, both new and used, are on sale, and a constant stream of buyers is looking for a good deal in hard times. I went there specifically to buy a copy of Eduardo Galeano's seminal work *The Open Veins of Latin America*.

During my 1,900 mile, 56-hour bus ride from Buenos Aires to Ushuaia, located at the southernmost tip of Argentina, I reread Galeano's thesis. Originally published in 1971, it is a scathing historical analysis of the factors that led to the development of Latin America before the early 1970s. His opening gambit sets the tone for the rest of the book:

> The division of labor among nations is that some specialize in winning and others in losing. Our part of the world, known today as Latin America, was precocious: it has specialized in losing ever since those remote times when Renaissance Europeans ventured across the ocean and buried their teeth in the throats of the Indian civilizations.[15]

As the bus continued south, stopping occasionally to pick up or drop off passengers in simple, sparsely populated towns, the gray, wintry, and unremarkable Patagonian landscape meant that I had sufficient time to contemplate Galeano's work and reflect on my journey. While most people in Latin America are struggling to make ends meet, across the region are innumerable examples of mediocrity and abuse of power: trials involving impeached heads of state, beneficiaries of *mensalão* (large monthly allowance for government officials in Brazil),[16] fraud and bribery, among others, seem to be normal, almost accepted as part of life in Latin America.

Many people in Latin America follow their own rules. Tax avoidance and evasion are common, particularly in cash economies, as are the sale, distribution, and purchase of pirated movies, music, clothes, and footwear. A poignant class divide further determines these informal rules: the rich have a wide range of options, from creative accounting involving multiple legal entities to siphoning money onshore to moving their funds offshore. The simple aim is not to share their money with the state, many of whose officials will take their cut through corruption, essentially converting them into servants of power.

It is as if the region hasn't fully recovered from colonial rule. Although the latter essentially was rebranded as independence two hundred years ago, creating an illusion of freedom following an alleged emancipation, the region continues to be run by a crony ruling elite, working shoulder to shoulder with imperialists overseas, and continuing to affirm Galeano's portrait of pillage. While the thesis of poverty, plunder of resources, market manipulation, violence, and repression holds true to different extents today, several differences seem to have emerged in the nearly half century that has elapsed since the book was written.

Major changes enacted on the environment, particularly deforestation, environmental pollution, water and land stress, have been game changers since the 1970s. Today, forestry, mining, banking, oil and gas, agriculture, and other industries pollute without any fear, remorse, or consequences. Governments have handed the captains of these industries carte blanche, which enables them to do as they please, much like Ayn Rand's hero, Howard Roark. Vulnerable people, plants, animals, and other forms of life simply have to acquiesce, comply, be injured, or perish—such are the immeasurable forces at work.

In terms of violence, state-sponsored brutality, including murders, still exists. Kidnappings and ransoms by criminal gangs are common across the region. Meanwhile, *secuestro express* and *secuestro relâmpago*, terms used in Spanish-speaking Latin America and Brazil respectively, refer to an express type of kidnap in which the victim is taken by force to an ATM (cash dispenser) in order to drain his bank account. Another scam, seemingly unique to Mexico, is to call someone pretending that they have kidnapped a family member, requesting an immediate transfer of bank accounts, or else. I have been told that the perpetrators run these scams out of prisons, where they have had mobile phones smuggled in.

More troubling still are extreme forms of violence: *El Pozolero*,[17] referring to the person who prepares the popular Mexican maize dish *pozole*, which normally has pieces of pork or chicken in Mexico, killed people by drowning them in acid; or *piñatas*, a wordplay on a traditional party game in which a papier-mâché figure is hung and beaten with a stick by children to release chocolates and candies, are also common, in which people are hung off bridges as a reminder to the general public and rival drug gangs. A new form of torture, which involves peeling the skin off of people's faces, demonstrates an even more brazen and sadistic means of inflicting pain on another human being. Violent initiation rituals for drug gangs in El Salvador include beatings to ensure that members are effective killing machines.[18] And Venezuela has widespread prison violence, which has led to a report in 2016 of cannibalism.[19]

These more extreme examples of violence, which are becoming more common across Latin America, act as important catalysts, exacerbated by severe and entrenched social inequalities, to encourage people to migrate to other countries (either more prosperous countries in the Global South, such as Mexico and Argentina, or the Global North, in particular the United States). The consequent effect is changing perceptions of space and how many people a country thinks it can accommodate. This, in turn, often brings with it negative consequences, such as racism and exploitation on arrival to their host countries.

Rather than try to confront these forces, the poor, who often are less prepared academically than the elite patriarchy of their country, find that leaving their troubled homeland and sending back money to their loved ones is a pragmatic solution to their plight, because they cannot afford to live in a gated community with armed guards to protect them. For the poor, the family replaces state welfare and provides an economic refuge because no other option is available.

These remittances are a staple in many Latin American economies, providing lifelines to families back home and to central banks, which look for ways to shore up their foreign currency reserves. The precarious agriculture industry, affected by deforestation, climate change, and landgrabbing, increasing levels of socioeconomic inequality and violence, together with the tenuous rule of law in numerous regions across the continent, have encouraged people to make the long journey north to pursue the American dream, sometimes at a cost for an illegal smuggler to help them cross the border of from US$6,000 to US$10,000 and, on some occasions, their lives.[20]

The king of remittances in Latin America is Mexico, which received US$36 billion in 2018, eclipsing income from oil, tourism, or any other industry, and making remittances risk-free passive income for the Mexican government. In 2018, the country's population was 126 million.[21] The failed drug war that started in 2006 cost more than one hundred thousand lives—mainly civilians.[22] Instead of providing their security forces with more adequate training and equipment to counter the parallel "narco state" that imports military weaponry from the United States, Enrique Peña Nieto's (2012–2018) government purchased lavish helicopters and planes,[23] adding to the Boeing 787 Dreamliner bequeathed to him by his predecessor, Felipe Calderón.

The real question is what incentive any greedy president would have to create conditions for people to stay. Knowing that a population is increasing means that if a few hundred thousand or even millions of people leave to send back hard currency to line the pockets of the privileged few and prop up the economy, this is viewed as better than if those same people stayed and contributed to a death toll that would further tarnish the ruling elite's image.

Brain drain is a parallel process. I have seen it many times with close friends who have completed master's or doctoral studies in the United States or Europe. Many do not have a compelling reason to return if they put their academic development, personal safety, and economic outlook high on their list of priorities. Mercifully, some who do return bring with them the necessary skills to help a new generation of Latinos learn and grow, as well as the courage and bravery to exchange the relative security of the Global North for the uncertainties of being back home.

Crucially, the population on the continent[24] has more than doubled since Galeano's book was published.[25] According to UN projections, the population of the Latin America and Caribbean region will increase to anywhere between 693 and 835 million by the year 2050.[26] More people means more mouths to feed, more houses to build, and increasing pressure to find ways for countries to make money in order to sustain these populations. In light of this, the aspects I witnessed on my journey across this continent gave me cause for concern because I can see the infrastructure cracking, security waning, and the environment collapsing.

The people whom I met, whether working class or middle class, were feeling the pressure, and, as their numbers have increased, so has the global population. One effect of this has been that the latter, fueled by both regional and global consumerism, has increased demands on beef, soybeans, fruit, sugar, wine, metals, coal, coffee, and more, which has, in turn, encouraged governments and businesses to improve the infrastructure across the continent.

Firstly, according to a workshop conducted by the IIRSA (Initiative for the Integration of the Infrastructure in South America), involving all twelve countries in South America, infrastructure across the region will gear up, with 531 transportation, energy, and communications projects requiring an estimated investment of approximately US$116 billion. These projects include new tunnels, railways, and bridges; the widening of highways and paving of nonexistent ones; the creation and upgrading of airports; and the upgrading and widening of ports.[27]

In theory, an improved infrastructure would benefit the people, because it would speed up transportation times—particularly important for perishable items—and reduce transportation costs over the long term, making products more competitive and increasing revenue. However, given acutely high rates of corruption in the region,[28] coupled with a long history of ecological disasters, it is impossible to predict exactly how much money will trickle down and the collateral effects on people and the environment once the project is finished.

An essential catalyst for these projects is access to global markets for foreign currency, loans, bonds, and other services. Some historical context of this phenomenon is described by Galeano: "The siphoning off of national resources into imperialist affiliates is largely explained by the recent proliferation of U.S. branch banks pushing up their heads throughout Latin America like mushrooms after rain."[29] For example, Chase Manhattan Bank acquired a number of local banks in the late 1960s, increasing its footprint in Latin America: 34 branches in Brazil, 42 in Peru, 120 in Colombia and Panama, and 24 in Honduras.[30]

These figures pale in comparison with banking in the twenty-first century. Today, one might add one or two zeros to these numbers, and include players from other regions. Citibank, Banco Santander, and other banks with headquarters in the United States and Europe have a huge retail presence in Latin America. Santander has more than fourteen hundred branches and eighteen million customers in Mexico alone.[31] During my stop in Buenos Aires, for the first time there I saw numerous Industrial and Commercial Bank of China (ICBC) banks, showing increased Chinese influence in the country.

Both in Latin America and beyond, low latency, high throughput trading systems—invisible and unknown to the general public—are changing the world, reducing humans and the environment to their essential utility. On any trading floor where I have worked around the world, traders keep a watchful eye on statistics, prices, and news: a bomb, change of presidency, the price of gold, unemployment figures, and so forth. Whatever data are received are later interpreted and transmuted into a binary positive or negative output, which will affect the price of whichever asset is traded, ultimately resulting in a buy or sell transaction.

These financial systems, and the aggressive practices that manipulate prices (chapter 1), paint a bleak picture. This is particularly the case when the latter are coupled with an existing physical infrastructure that is continually being upgraded, albeit slowly and incrementally, not to mention the ambitious plans the IIRSA outlined, together with unscrupulous custodians of power.

Stealthy banking systems and overt expansionist plans are not alone in seeking people's utility. This search also emanates from the individual, fueled by popular culture. So, instead of having deep conversations about what is going on around them, a reductive view of the self and the other is embraced, with a focus on extrinsic values and, ultimately, the other's utility or worth, often overlooking the true value and beauty of other people and the natural world.

It makes more sense to nurture the ego and forget about the real issues facing our time: cocaine abuse, cosmetic surgery, and dumbed-down soap operas serve as a major set of distractions, and are all very popular in the Latin American region. Cosmetic surgery exemplifies how important image is in a region riddled with poverty, and it helps to cement the disconnect between empathy and reality. Behind the facade of musicians, actors, politicians, footballers, and the other beautiful people who grace the television, newspapers, and social media, the potential of this continent is being hoarded among the powerful few and their elite, often historically connected, ties overseas.

In spite of the innumerable pressures the people I met on my journey have experienced, they continue to strive, trying to provide for their families, communities, and the environment. Many were taking things a step further and looking for immediate solutions. Their good hearts and strong characters had adapted to the theft, violence, political and financial instability, as well as climate change, but how much longer will they be able to withstand such gargantuan amounts of stress and uncertainty? And what price, if any, could be put on their suffering?

After a quick stint in southern Chile, which included crossing the Strait of Magellan to reach Tierra del Fuego, I finally reached Ushuaia, known to Argentines as the city "at the end of the world," referring to the city located at the southernmost latitude on Earth. I was less than seven hundred miles from Antarctica in the middle of winter, and yet other guests at the hostel went hiking instead of skiing and snowboarding because it hadn't snowed in the past week. I lay on my top bunk, shell-shocked, beginning to digest what I had seen in this wonderful continent.

II

UNITED STATES AND EUROPE

· 6 ·

American Power

ELECTRIC VERSUS GASOLINE CARS

Having spent such a long time traveling on my own in Latin America, I was delighted that my good friend Madison Koen, a violinist from Texas, joined me in Miami at the start of my journey along the East Coast. Neither of us knew this part of the United States, so it was an equally new and thrilling trip that added more memories to our friendship. We stayed at a friend's apartment in Coconut Grove, a lovely neighborhood with lots of fancy cafés, bars, restaurants, and marinas, and spent the first couple of days exploring the city.

Among the first things I noticed during our long walks was how empty the sidewalks were and how many people were driving Ferraris, Porsches, and Mercedes, all of which seemed to fit very well in a landscape of abundance and luxury. It was something I saw at this scale for the first time since leaving Mexico City nearly a year earlier, and something that would repeat itself across the Global North, right through to Hiroshima (chapter 17).

Being a former car enthusiast, I was aware that several of the luxury car brands had started to produce exclusive models employing hybrid technology. These combine a high-performance gasoline engine with battery-powered electric motors, taking their supercars into hypercar territory and creating a new range of production automobiles that can do 0–60mph in less than three seconds, something almost unimaginable a decade ago.

Meanwhile, Tesla, spearheaded by the forthright and outspoken Elon Musk, hit the headlines for its ability to parallel supercar performance figures in its all-electric sedan, the Model S P100D,[1] in 2016. That a Californian manufacturer could build a car that can hit 60 miles per hour in 2.5 seconds at less than a fraction of the price of European hybrid hyper-cars demonstrated something completely out of the ordinary. I saw many more Teslas than I expected in Miami, and I was surprised when they drove past because they were as quiet as a golf cart. I felt that I had taken a leap forward into the future.

Nevertheless, regardless of this apparent leap, I couldn't avoid the fact that automobiles are very inefficient, regardless of engine type. Lots of CO_2 is used in their manufacture due to a global supply chain that sources different raw materials and assembles parts in multiple locations. In the case of hybrid cars, just as with the bioethanol additive used to reduce CO_2 emissions at the tailpipe in Brazil (chapter 3), they still burn fossil fuels, which provides the first stumbling block to the utopian dream.

Furthermore, the battery power used in both the all-electric and hybrid automobiles produces several problems. Although all electric automobiles do not have a tailpipe from which to spit out particulate matter, CO_2, and other harmful gases, the energy grid that provides the energy to begin with usually has a carbon footprint.

In the case of Florida, all-electric automobiles produce slightly less than five thousand pounds of CO_2 equivalent, compared with about eleven thousand pounds of CO_2 equivalent for gasoline engines. One could argue that this still brings a huge benefit, given the more than 50 percent reduction in CO_2 emissions from such electric vehicles. Nevertheless, this is where the next problem begins. Many car batteries used today are lithium-ion, which are relatively light and offer good performance; however, lithium extraction is notoriously challenging.

In buying electric vehicles, the consumer has simply transferred the risk elsewhere. Manufacturers are all too keen to brush the facts, which they are well aware of, under the carpet. The chief exporters of lithium today are Chile, Argentina, and Australia. Manufacturers will continue to exchange hard currency for metals without problems because, as we have already seen, foreign companies that work in Latin America can get away with polluting in the name of shareholder value, encouraged by lax environmental laws and/or the officials that they have bribed.

Is this likely to suggest that although lithium reserves are exhausted and environments in distant lands are being pillaged and polluted, affluent cities may breathe cleaner air by adopting this technology across the board? Of course, it's not so simple. Another dark side of electric cars is the cobalt their complex batteries need. A lawsuit against numerous high-profile American companies including Tesla and Apple, filed by International Rights Advocates in Washington, D.C., documents several severe injuries that resulted from exploitative child labor during extraction of cobalt in the Democratic Republic of Congo (DRC).[2] Knowing this, and other transgressions against human rights and the environment, later would lead me to interview miners in South Africa (chapter 14) after I was unable to obtain a visa for the DRC (a notoriously difficult procedure).

The consumer in the Global North is unlikely to be aware of these points and buys the hybrid or all-electric automobile in good faith, unaware that the metal extraction done far away is, in fact, problematic. Manufacturers encourage such consumption on the false precept that the vehicles have zero emissions, thereby replacing any guilt over buying gasoline- or diesel-powered vehicles with a feeling that their decision will make the world a better place.

Such an erroneous conclusion could also be seen at the expensive grocery stores with organic food, the cafés with fair-trade coffee from faraway lands, and the generally laid-back feeling of the city of Miami as a whole, which created the aura of being somewhere that wasn't so ecologically unfriendly; yet, I could see many people driving large, heavily polluting vehicles, often on their own. In spite of this incongruence, I really liked Miami and was pleasantly surprised to walk into a restaurant to be greeted with the Spanish "buenas noches" (good evening), even though I hadn't uttered a word.

After our meal, we walked outside, and I could see even more luxurious and exotic sports cars. The global car industry presses the gas pedal unrelentingly, promising the illusive dreams of freedom, adventure, practicality, and safety, when in fact, car ownership today is quite the opposite: being stuck in traffic jams and debt while choking on collective exhaust fumes can hardly be called a liberating experience, even if the car is considered to be a luxury one and the convertible hood has been lowered.

But what is the true cost of oil, economically, environmentally, and with respect to human rights? The first part of the question is easier to answer in a global economy that is extremely sensitive to oil prices given that so many industries depend on it, including the transportation, pharmaceutical, chemical, and energy industries.

The 1970s oil crisis clearly demonstrated the interconnectedness of economies. A quadrupling of oil prices helped to cause a stock market crash, interest rate cuts by central banks, a recession, and a temporary reevaluation of energy security. At the time, President Jimmy Carter asserted, "By the end of this century, I want our nation to derive 20 percent of all the energy we use from the sun."[3] Keen to lead from the front, he installed solar panels on the roof of the White House, only to have them taken down by President Ronald Reagan (when oil prices had gone down again). And there I was, in Miami, endowed with beautiful, sunny weather nearly forty years after President Carter's bold and ambitious idea, wondering how such a dynamic and cosmopolitan city could still be stuck in the past.

The general consensus is that maintaining low, stable oil prices is good for the global economy. Oil, quite literally, runs the world's engine. It ensures that the goods that pass through the Panama Canal (chapter 1) and the other important shipping lanes around the world are as cheap as possible, that we can fill our automobiles with gasoline without breaking the bank and gain access to affordable pharmaceutical drugs, not to mention all the plastic bags, straws, cups, and other oil-derived products given away for free, reflecting the low price of the raw material.

Besides the obvious impacts on human health that burning hydrocarbons causes, numerous environmental implications range from global warming to oil spills. Thus, the true environmental impact of oil, which permeates so many sectors, becomes a quagmire. The first part relates to consumers. We don't even know where the oil comes from, nor are we told what the real CO_2 emissions are from well to tailpipe, meaning that the figures car companies produce, even if they are accurate, fall short of demonstrating the true amount of oil used to power each car. This is particularly true because such figures do not take into account extraction, refining, and transportation of the fuel, nor is the carbon footprint related to its manufacture factored in (e.g., companies should be calculating the carbon footprint for the assembly of the car, then dividing it by a certain number of miles—that is, the expected lifetime of the car—and adding that to the grams per kilometer figure they publish. If they did this, we'd have a better idea of what we were emitting).

To compound matters further, outright lies relating to emissions were exposed during the Volkswagen[4] and Mitsubishi[5] scandals in 2015 and 2016, respectively. One needs to exercise caution when putting faith in official figures, not only because of a risk that the manufacturer might be deliberately understating figures, but also that the figures have been created under conditions that will almost never be reproduced in the real world. Given the millions of cars, lorries, and buses on the road, and thousands of planes in the sky at any given moment, this consideration should not be underestimated.

For all these reasons, together with the fact that Miami is a coastal city and, therefore, acutely at risk from rising sea levels caused by climate change,[6] I was concerned by the lack of progress I saw there. The luxury hybrid and electric vehicles implied a transition in the right direction. But the sheer volume of large, extremely powerful gasoline- and diesel-engine vehicles made me think that perhaps some people may be unconcerned by, or unaware of, the environmental and human rights issues related to the fuel (chapter 9) they burn and don't feel compelled to ride a bike or use public transportation that, for the most part, was empty.

SOLAR-POWERED ROOFTOPS

A few months after my visit to Colombia (chapter 2) where I had looked into solar-powered water heaters followed by a visit to the community affected by the Glencore open-cut coal mine, a report by the International Energy Agency (IEA) stated that renewable energy production had overtaken coal.[7] From the moment I arrived in Florida to the moment I left Boston and flew to my hometown of London, I kept searching for examples of solar power and how we might overcome our dependence on fossil fuels, because what I had seen and read began to trouble me.

Soon after I arrived in Miami, I passed by an IKEA store with solar panels on its roof measuring tens of thousands of square feet. That put a smile on my face. Sadly, my enthusiasm would be short lived: not a single large energy company accepted an interview request. This came as a great surprise because generally most companies, institutions, and individuals that I approached during my journey were very helpful and forthcoming when they learned about the project's intentions. In contrast, the energy companies I contacted seemed surprisingly coy about engaging in a dialogue, despite having so much power and influence over people's lives. Following these setbacks, I spent a couple of days walking around Miami to see if I could find solar panels in the flesh. Alas, I didn't find any besides the ones above the IKEA store that I had seen at the start of the trip.

Before leaving the city, I met with Maria Ceron, an energy analyst, to gain further insight into energy trends in the United States. She had this to say: "First of all, we're seeing a kind of leap-frog in technologies, in a similar way to what we saw a few years ago with mobile phones. Before we used to use fixed line phones, and today we use mobile phones. A similar thing is happening in the energy sector."[8] In California, people are installing their own solar panels. In Maria's view, that empowers consumers to make a difference through their own actions. She also mentioned that other states, such as New Jersey and New York, were catching up.

Maria believes that solar power has huge potential not only in the continental United States, but also nearby on islands such as Puerto Rico and the Virgin Islands, which could be good test beds for this renewable source of energy. Speaking about some of the obstacles to solar energy, Maria observed, "There are always monopolies, there are always many interests."[9] A possible reason why I didn't see much in the way of solar-powered solutions in Miami could be related to a large nuclear power plant, Turkey Point, which, Maria explained, accounted for "very cheap" energy in Miami. We also talked about another part of the world with huge solar potential: Africa. However, a major inhibitor she sees there is investor confidence.[10]

Our meeting happened just a few weeks after the CO_2 level passed four hundred parts per million (ppm),[11] an inauspicious milestone that serves as a good indicator of global warming.[12] "As a woman, as a scientist, and human being," Maria commented, "I ask whether I should bring a child into the world."[13] I, too, faced this ethical quandary, and at that time had made a conscious decision to avoid procreation (although that would change later—a choice based more on love than logic).

Maria spoke about the transcendence of watering plants in the garden or riding a bike, both of which would become important things in my life as the whole journey unfolded. Her view that "we have a natural connection with the environment"[14] is one I share. Although the political will and market conditions to reverse climate change may not be there, we can all make a difference in the things that we do.

Although I left Miami with more questions than answers, I was grateful that I had met someone who understood and cared about the planet as much, if not more, than I did. Luckily for me, some doubts related to energy would be resolved as I headed north toward New York City, and later in Copenhagen, when I would have an inspiring meeting about wind energy and witness remarkable examples in action (chapter 8). As Madison and I arrived at the newly built Greyhound bus station, I felt that I had been shielded from the inequality I knew exists in the United States. Everything to that point looked great and luxurious. Appearances can be deceptive.

In Atlanta, I finally managed to see several examples of solar-power installations. One was in the south of the city, where I visited the seemingly disused landfill in Hickory Ridge. Partially visible from the edge of the freeway, solid waste protrudes from the ground and is covered by a green canopy that has solar panels attached to it. A kind of two-for-one project, the landfill is designed to collect methane generated from the decomposing materials[15] in a similar way that Fabio did in the favela in São Paulo (chapter 4). The gas can be collected and used as fuel, with electricity generated from the solar panels.

Regrettably, I was advised that the project manager was unavailable for an interview while I was in Atlanta. However, I had the opportunity to discuss what I saw with a mechanical engineer who knew about the site: "If you look at the energy costs used in its production and compare it with the energy gains following its implementation, you'll find that it's completely inefficient. It's just a gimmick." Such a view would imply that the project was a form of greenwashing used to manipulate the well-intentioned but uninformed consumer, who might see this as a step forward when, quite possibly, the reverse is true.

Following my visit to the landfill, I went to Stanton Park, a state-of-the-art installation with a set of solar-paneled roofs that can power the modest electrical needs of the energy efficient—and I suspect very expensive—LED lighting system and public lavatories. I spent several days searching for more examples in the city, the lack of which began to trouble me. I had a nagging feeling about why states such as California seemed to be far ahead of Florida and Atlanta. To gain greater insight, I contacted Aaron Scranton, a certified solar consultant, who identified two major inhibitors.

The first is policy, which he believed was the biggest challenge to solar implementation in the United States. This, he explained in an e-mail, was related to the low cost of subsidized traditional electricity sources: "There is a longstanding relationship between supply in the coal industry and government subsidized nuclear plants that have kept the cost of traditional energy very low on the consumer end. This has had the effect of insulating the market at every layer."[16]

A constant, reliable, and cheap supply of conventional natural gas and coal, as well as shale gas (through fracking), helps keep costs down. Another important consideration that reduces the potential of solar power is mass energy storage, a perennial problem for industrial-scale electrical systems, and beyond the capacity of lithium-ion batteries. The conundrum is that it is very difficult to store an oversupply of electricity from one source. Aaron told me that numerous technologies are being tested, and we are close to a tipping point to solving this problem.[17]

I also wanted to understand Aaron's view about perceptions at an individual level. He feels that things are changing. Consumers are becoming less skeptical and more savvy. Through constant exposure to climate change, people's awareness of what they are doing and what can be done is increasing: "It's also just a matter of having these systems reach a certain level of saturation where the public sees them in use consistently and begin to wonder what they're missing out on."[18] Aaron considers solar power the obvious end to the chapter on traditional energy systems and infrastructure, with this caveat: "Unfortunately, the lobbyists will see to it that it's stretched out as long as profitably possible."[19]

That Miami is heavily dependent on nuclear, gas, and coal power, with such a large number of gasoline-powered cars and few solar-powered initiatives, when it has the potential to be a leader in the field, suggests that the old-school, oil-hungry patriarchy that has so much power in the United States (with close ties to Saudi Arabia and elsewhere in the Middle East, as well as Russia), doesn't have a vested interest in things changing for the better. A few months after my journey along the East Coast, Rex Tillerson, a former executive of Exxon, one of the world's largest oil companies, became U.S. secretary of state[20]—a further reminder of the center-stage place fossil fuels occupy. Thankfully, as the remainder of my journey across the United States would demonstrate, hope was still present, waiting in the wings.

I never used to believe in fate or destiny. Six years of science at university had weaned me off such romantic ideas. As I walked through the parking lot of Whole Foods in Brooklyn, New York City, about to step into the store to buy some lunch, I saw a BMW i3, a new all-electric vehicle, park at the electric point adjacent to the store's solar panels. Precisely when I was looking for more examples, I asked myself if this could get any better. I approached the driver, Steve Nelson, and asked him a few questions about the BMW.

Unable to decline Steve's opportune and somewhat fortuitous invitation to have lunch together, we bought food from Whole Foods and ate at Steve's office located in a warehouse a few blocks away. It seemed to be the perfect place for the headquarters of an ecologically driven NYC-based start-up company. Steve led me out to the terrace, where I saw a demonstration model: a stylish and elegant aluminium canopy with lightweight struts about nine feet off the ground that created a roof structure, at the same time complying with city fire department regulations and allowing access to emergency services.

As a solar consultant, part of Steve's job is to explain the tax incentives and credits designed to get customers on board. "People are motivated for a lot of different reasons, ranging from concerns about where their energy comes from through to the economics of it."[21] For some customers, installing a solar roof now is seen as some sort of tech toy, but Steve believes they will be ubiquitous in the future.

In addition, the financial benefits make solar-paneled roofs an attractive proposition. The break-even rate—that is, when the client makes a return on the investment—depends on whether the panels are acquired as a loan or purchase, the unit cost, and, of course, how much the client saves each month on the utility bill. Joseph Lipari, solar operations manager, explained a common situation that clients experience when their meters spin backward: "They're watching their electricity meters send electricity back into the grid."[22]

In 2016, a standard installation of eighteen panels cost about US$35,000 before tax incentives. Although the final price after discounts is still likely to be five figures, homeowners get an additional benefit that comes with a twenty-five-year standard warranty: the solar panels add value to their home and can help resell the property, as the next buyer doesn't have to go through the installation process and automatically has lower energy bills, in addition to the privilege of reducing their carbon footprint.

I wanted to learn about the panels' performance during suboptimal weather conditions, such as wintertime, when daylight hours are significantly fewer because of the city's high latitude, and when they include excessive cloud cover or snow. Before Steve does an installation, he takes readings on the rooftop, including a shading analysis, and factors in shorter days and snow coverage to estimate the production of a given canopy configuration throughout the year.

The impromptu meeting with the team left me feeling more optimistic after my experiences in Miami and Atlanta. At that point, Brooklyn Solarworks had been in operation with the current canopy for two years and averaged two new customers a week, positioning itself to tap into a market that is growing exponentially. By getting involved in the transition from traditional energy sources to solar, the team will contribute positively toward the city administration's target of 1,000 MW by 2030, approximately twelve times the current capacity and constituting enough to power more than 250,000 homes,[23] making a significant contribution to reducing carbon emissions. Over the coming days, more examples of solar panels in New York City began to emerge. From the rooftops of warehouses in Gowanus in Brooklyn, the Citibank bike stations all across the city, to random, stand-alone installations above private homes, I began to see the city in a new light.

BIKES, BUSES, AND PLANES

As I traveled the world, both on this and previous trips, I witnessed the most horrendous traffic jams, from Nairobi to São Paulo, with the halo of pollution shrouding each city a common icing on the proverbial cake. Although I didn't come across any bike-share schemes in Miami, I did see some in Atlanta and New York City. They provided a ray of light, but the mismatch between the small numbers of bikes and the enormous volume of automobiles diminished my jubilation. That being said, as I would discover later in my journey in Europe (chapter 8), some cities have already started to wean themselves off hydrocarbons and have invested heavily in bike infrastructure; one even has a zero emissions target. For now, in the United States, public transportation was the closest to eco-transit that I could find routinely.

The long trips on Greyhound, America's largest bus network, highlighted some public transportation challenges in the United States and elsewhere. The bus, when compared with the Scania and Mercedes models I had traveled on many times in Central and South America, seemed archaic despite retrofitted electrical outlets to charge my phone and the provision of free WiFi. The rock-hard seats made it impossible to get any decent rest, and perhaps most surprising was the lavatory at the back of the bus: an open pit toilet with no hand basin, soap, water, toilet paper, or any semblance of hygiene.

Is this really the way human beings in the world's richest nation should be transported? I was immediately reminded of Hemingway's *To Have and Have Not*,[24] and the sad idiosyncrasy of such a beautiful and powerful nation: the haves and have nots. The fact that fifty million people in the United States live in poverty,[25] equal to the population of Spain, meant big demand for relatively cheap transportation that I, and the other largely black and Latino passengers, could find.

Here's an example emblematic of how cheap hydrocarbons were during my journey: a Norwegian Air flight from Boston to London on a brand-new Boeing 787 Dreamliner cost US$125, meaning that the price to travel at six hundred miles an hour was less than four cents per mile. Can anyone imagine taking a taxi, or any form of transportation cheaper than this? Air travel in Europe has become so inexpensive that it is cheaper to fly than to take the train. In an article published by the BBC, titled "Blogger Travels from Sheffield to Essex via Berlin to Save Cash,"[26] the self-evident absurdity of traveling longer distances to save money clearly shows that it is cheaper to destroy the planet than to be ecologically conscious.

However, the airline industry is facing a turbulent period given the effects of coronavirus, a storm I would enter toward the end of my journey in Asia (chapter 16), which has caused a dramatic fall in passenger flights and a steep rise in ticket prices for some routes. Although it's impossible to foresee how long social distancing and other measures to reduce the spread of coronavirus will go on, the extractive industries still press forward.

It took millions of years for dead plants, trees, and other vegetation to turn into coal, gas, and oil, and since we have mastered its extraction, we have also perfected the art of destruction. At any given moment thousands of planes are flying dozens of thousands of people to business meetings, holidays, home, and so forth, and simultaneously, we're experiencing oil spills, habitat losses, pollution, and acidification of the air and sea.

Mercifully, it would appear that some nations are paying attention: *Flygskam*, meaning flight shame, is a newly created concept in Swedish, which, it is believed, has had a positive effect through a 4 percent reduction in flights.[27] Many of my friends and I travel around Europe by train even when it's more expensive than flying. Our consciousness is awakening slowly. Today, more people no longer feel able to accept silently that consumerism out of control should be part of our reality in the twenty-first century: pragmatism is becoming more difficult and less tolerable as we slowly see costs increasing at economic, ecological, and emotional levels. In evaluating what is happening in our own worlds and imagining how this is connected with people and the environment on the other side of the planet, we slowly begin to understand a shared destiny.

· 7 ·

Food in the City

MY FIRST THANKSGIVING

While checking out of the guesthouse in Atlanta, where I had been exploring solar power, the owner, a black woman in her fifties, asked where our next stop was. When I told her that we had booked an overnight Greyhound bus trip to New York City, she insisted that my friend Madison—who was traveling with me during the East Coast leg of the trip—and I should stay for Thanksgiving. We accepted and asked how we could help. The owner said they had everything except dessert. Naturally, we obliged.

The owner prepared a Thanksgiving lunch with a twist because her daughter was vegan: green bean casserole, corn bread, macaroni and cheese, stuffing, tofu turkey, cranberry sauce, twice-baked potato casserole, and a vegan carrot cake that Madison and I brought. I felt both humbled and honored by our host's generosity of spirit for letting two strangers to join her and her family for this modern take on a traditional meal to mark an annual celebration dating back to the seventeenth century when colonial pilgrims shared a feast with the Pokanokets, an indigenous tribe based in present-day Rhode Island.[1]

Not wanting to outstay our welcome after dinner, we went to a Walmart a couple of miles away (the only store open at that time) and stocked up on a few snacks and sodas ahead of our bus trip. While we sat outside in the cold eating sale-priced vanilla ice cream, four police cars arrived and arrested a black man who was shouting loudly, in distress; it wasn't clear what, if anything, he had done. Then I reflected on the depressed economic conditions in the United States that have

85

affected so many people, mostly black men, whom I saw living on the streets of Atlanta. I had witnessed similar scenes all too often during my journey across Latin America.

Following our second overnight Greyhound ride, we arrived in New York City on Black Friday, a day of massive discounts immediately after Thanksgiving. Unlike the man arrested outside the Walmart, presumably because he needed to consume out of desperation, the desperate consumption by the masses had purchased more than US$3 billion in goods that day. Interestingly, midtown Manhattan was empty, so I tried to figure out where the shoppers were. Perhaps they were ordering their sale-priced goods online because it's more comfortable, time efficient, and eliminates the risk of being crushed in the stampede of people eager to grab a bargain.

It was painful to imagine that people might derive so much joy from buying goods for the sake of buying, rather than have a more reflective, genuine, joyful celebration. I pondered on the ecologically, ethically inspired vegan Thanksgiving lunch that I had shared with a black family and Madison, a bilingual English-Spanish mariachi violinist who sees the world through the lenses of knowledge, compassion, and respect in the work that he does in different communities in the United States and Mexico. The meal was a wonderful experience, which placated some fears I had about unbridled senseless consumerism. The sense of hope it brought would be revived as I explored New York's sustainability scene.

NYC HERBS AND GREENS

At the other end of the consumerism spectrum, I met dozens of people throughout my journey committed to making not only their local environments but also the global one more pleasant, healthy, and sustainable. An integral part of the changes I witnessed was happening in the food industry. Above the Whole Foods store in Brooklyn, I met with Julie McMahon, a manager at Gotham Greens, a leading producer of hyper-local[2] food in several American cities.

"This is the first commercial scale rooftop greenhouse integrated into a grocery store in the US,"[3] said Julie. As we stood in the twenty-thousand-square-foot greenhouse, I looked across a sea of intense dark colors of leafy greens, basil, and other herbs. All of the seeds were non-GMO. Instead of using pesticides, they use insect predators as an effective natural alternative to the horrendous chemicals that, for example, I would see in action in Kenya (chapter 12).

During the tour of the Gotham Greens urban farm, Julie explained the company's aim to reduce greenhouse gas emissions. An important step toward this goal is to avoid the waste of around 30 percent of basil that is farmed out of state. By growing, producing, and distributing hyper-local basil both to retail customers and restaurants, they can get fresher, more nutritious basil to their clients and reduce the overall carbon footprint since the basil doesn't have to travel hundreds of miles. That also means less food waste. This same principle applies to the other herbs and leafy greens they produce.[4]

Some of the energy the farm uses comes from wind turbines and solar panels. By harnessing renewable energy and using their experience gained from multiple sites, including an even larger, one hundred thousand square-foot urban farm in Chicago, and combining these with strategic partnerships such as their downstairs neighbor in Whole Foods, Gotham Greens can consistently reduce the time from farm to fork and provide food year-round.

Bell Book & Candle, a mid-priced restaurant in Manhattan's West Village, features a simple and healthy menu that includes a Rooftop Mixed Greens Salad that I sampled. The greens come from the restaurant's rooftop garden. But not everyone can afford such good food as that carefully nurtured and provided by this eco-conscious establishment. Given the purchasing power of major currencies such as the U.S. dollar, sterling, and euro, many quality foods, wines, and other beverages are imported to the United States and other countries in the Global North. Avocados may come from Mexico and organic bananas from Peru, a kiwi may cross the Pacific Ocean from New Zealand, spring water from Fiji, coffee from Congo or Colombia, and so on. All essentially are luxury items accessible only by the few, not the many, made possible by the international shipping industry (chapter 1).

In order to facilitate the high-volume trade of staple products at low prices for the mass market, which lies at the other end of the spectrum, the same shipping infrastructure moves large quantities of staple ingredients thousands of miles, leaving a trail of destruction in their wake. The very fact that the products have traveled great distances adds to each product's carbon footprint, not to mention adversely affects communities, as I had found to be the case with the dangerous, ruthless soybean industry in Argentina (explored in chapter 5). In another instance, wildlife such as orangutans are disappearing in Indonesia because of logging to make way for palm oil plantations[5] in exchange for palm oil that makes its way into anything from toothpaste to pizza.[6] Collectively, these destructive patterns show how inextricably linked consumerism is with environmental and human rights violations.

This puts the consumer in "zugzwang," a German loan word mainly used in chess when a player is set to be at a disadvantage regardless of the next move. Consumer choices are extremely difficult due to nuances in the harvesting, manufacture, packaging, distribution, and sales processes, all of which have ethical, environmental, and economic considerations. This is particularly the case when it comes to food, because the soil, seeds, and pest control may or may not meet organic standards. Meanwhile, the farm-to-fork trajectory may involve one or several journeys in a car, bus, ship, and/or van. In some absurd cases, the food (such as a banana or an orange) may come in plastic packaging. Finally, working conditions at every step in the process, often hidden from consumers, have important ethical considerations.

All along this chain, people are making money, some more than others. Some are exploited; others walk away rich. And yet, in the end, all the consumer wants is the product. The choices are bewildering, especially for those who have a greater degree of awareness and are willing to enquire. Is it better to buy GMO bread, made oppressively next door, that doesn't involve chopping down a rain forest, or buy fair trade organic food that traveled five thousand miles and required a rain forest to be cleared for it to be produced and oil from an oppressive state burned in order to transport it, polluting the environment in the process?

Although we never face such obvious binary questions, we are presented with choices that have a paucity of data. So vast are the unknowns that we essentially make educated guesses at best, and blind ones at worst. Almost every decision has a saving grace and a downside. For an increasing number of people, the decision simply is based on affordability, bringing us back to the reality of unequal access to nutritious food that has the lowest possible impact on the environment.

During times of financial hardship, cash-strapped individuals naturally are drawn to low prices on chicken, pizzas, hamburgers, and other fast food. Low-income families, including children, sometimes consume cheap high-calorie meals to stave off hunger,[7] but at a price: many of these foods contain high levels of saturated fat, sodium, and sugar, all of which are widely known to have negative impacts on human health.

According to the National Institute of Diabetes and Digestive and Kidney Diseases, more than 37 percent of adults in the United States are considered obese; for children and young adults between two and nineteen years of age, the figure is approximately one in six.[8] The same institute lists type 2 diabetes and heart disease, among other conditions, as being health risks associated with being overweight or obese.

Of course, sick people need health care and lots of medicine—anti-hypertensives, antiglycemics, and statins (for cholesterol)—not to mention having increased risk for dialysis and other kidney issues that require medical attention and, therefore, someone else to make money. Every problem has a solution, but surely it makes sense to avoid the problem in the first place. Or is it better to have millions of sick people quite literally eating the planet toward disaster, where the only winners are the shareholders of the agrochemical, pharmaceutical, food companies, and other industries that profit from people's excessive eating habits?

The notion that the planet has too many people is contentious because the range of consumption of natural resources (water, hydrocarbons, energy, raw materials) and purchasing power is so extreme. In any case, addressing obesity would be a giant leap forward to save lives, reduce burdens on health-care systems, reduce pressure on ecosystems, and protect the rights of vulnerable peasant farming communities. This, in turn, raises questions about social inequalities, health education, and legislation, and how careful coordination of these elements might have a positive impact by modifying the conditions that are leading too many people to premature death.

Back at the Greyhound station in Manhattan, I saw numerous vending machines filled with soda, chocolates, cookies, and potato chips. I would be hard pressed to say they contained anything healthy besides spring water. I sensed that I had witnessed the creation of a food elite, both in terms of knowledge of the topic, and purchasing power, which meant that people were being left behind. Even in 2016, when the United States was not in a recession, I saw numerous independent food outlets offering dollar pizza slices and huge soda drinks for just a few dollars. Supply and demand rule the world.

That New York City had several restaurants of the same ilk as Bell Book & Candle, as well as rooftop farms, demonstrates an important step forward. But I was cautious about celebrating an early victory: many of the more ecologically oriented restaurants and food outlets I visited were not everyday options because they are expensive. However, when I learned that the Michigan Urban Farming Initiative in Detroit had converted an inner-city space that provides healthy, organic food for two thousand households "at no cost to the recipients,"[9] I felt a bit more encouraged.

The potential to modify spaces and concepts to change our own lives and those of others for the better is staggering. On one hand, the United States taught me about the paradox of poverty living within a bubble of opulence, and coexisting with the contradiction of the greatest number of Nobel Prize winners living in the same space as those who harm themselves through ill-informed eating habits. On the other hand, my journey there highlighted to me that the challenge is to educate people and provide ways for them to gain access to nutritious and affordable foods with lower carbon footprints while being kind to the natural world and people. The solutions and the knowledge are all there. The next step is use them to make it happen.

A TOAST TO REDUCING FOOD WASTE

I first learned about Tristram Stuart's food activism through his organization, Feedback, which created an event in London called Feeding the 5,000, using food waste to feed those in need.[10,11] Following widespread support, this was then re-created in New York City, Washington, D.C., Brighton, Milan, Athens, and Paris. In common, all of them suffer, to a greater or lesser extent, from two of the greatest ills we see today: food waste and poverty.

Notwithstanding the worldly limitation of converting five fish and two loaves of bread to sustain five thousand men, besides women and children, this initiative highlights a tragic contradiction that exists in the United Kingdom, the sixth-richest nation on the planet: excess and dearth silently coexist as unlikely bedfellows in a house of greed and benevolence. The same was true in many of the cities I had visited at that point, both in the Global North, such as Atlanta and New York City, and far too many in the Global South.

"Humans are now a majority urban species. It means that most of us live further and further away from where food is being produced, and we are losing touch with everything that goes into making food," said Tristram.[12] Today, more than a third of suitable land is used for agriculture. That is putting additional strains on existing ecosystems and pushing humanity toward a precipice, an avoidable situation that, as Tristram put it, "we are currently eating our way to."[13]

Our conversation moved to narrowly conceived ideas of perfection, "produced by discarding 30 percent or 40 percent of the products that the earth produces because they don't look right." Tristram was of the opinion that supermarkets have helped to create the illusion of abundance, together with uniformity, to create a "perfect visual display of food" resembling some kind of ornament rather than an essential product.[14] Supposedly ugly vegetables are thrown away because they don't look as perfectly formed as the images that have been etched into society's mind.

In disregarding this food, not only has the whole supply chain process been wasted and, with it, the carbon footprint increased, but the consumer essentially has shown intolerance to anatomical variations that nature brings. I am unaware that a single carrot with two roots, a zucchini with a prominent bend in the middle, or an asymmetrical apple have any more or less nutritional value than more common variants of the same species. Yet, if they were on display, many people probably would elect a more commonly shaped food item.

Among Tristram's range of projects within the field of food waste is the recently launched Toast Ale, a premium ale that includes the equivalent of a slice of toast in every pint. Almost immediately opposite the Supreme Court of Justice in central London lies the small, unassuming Essex Street Brewery, which produces Toast Ale, in the basement of a building where I met with Vanesa de Blas, head brewer. This bubbly, energetic lady, whose previous training was in the wine-making business in Spain, showed me the large vats where the ale is brewed using toast made from fresh bread that bakeries otherwise would throw away.[15]

The process follows four simple steps: malt and the toasted bread are mixed with hot water for an hour, then the liquid is extracted and boiled for an hour and a half with bittering hops, releasing antioxidants into the beer and thus avoiding the need for artificial preservatives. After boiling, more hops are added, and the mixture is boiled for a further five minutes, allowing the aromas to be preserved, and giving the beer its characteristic hoppy and citric flavors. The penultimate step is to pour the liquid into the fermentation tank and mix it with yeast, then let the beer ferment for a week. Finally, Vanesa applies a natural carbonation technique, similar to that used in champagne, to create the finished product.

Although most of the beer is bottled and sold across the United Kingdom, including most recently in the prestigious Selfridges store on London's Oxford Street, some of the beer is on draft. Not wanting to be a bad guest, I had an uncharacteristic morning tasting session of the two draft versions of Toast Ale. The pale ale had a large creamy head and slightly acidic taste, both of which are characteristic of similar ales, while the dark ale had chocolate notes with a typically bitter stout aftertaste. "I think that if you were going to do a blind tasting, it does taste similar to a regular pale ale," said Vanesa. The additional wheat from the bread helped create the lovely creamy head on my beer. "They [our customers] think that it's a fantastic idea. It's like a guilt-free pint. Do you know what I mean?"[16]

Tristram's drive and commitment to Toast Ale and the Feeding the 5,000 project reflect much-needed ambitious initiatives, but it is shameful that an individual should need to ensure basic sustenance for citizens who live in one of the richest nations on earth. People are falling through societal cracks that have become gaping holes, as is the amount of food being deliberately wasted, meaning that food banks are on the rise in the United Kingdom. Trussell Trust Foodbanks, whose vision is simple—"To end hunger and poverty in the UK"[17]—provided more than 1.5 million three-day emergency food packages to people in need in the 2018–2019 UK financial year,[18] 19 percent higher than the previous twelve-month period.[19]

This strange contradiction isn't unique to the United Kingdom. Approximately eighty-eight million tons of food is wasted annually across the European Union,[20] and, yet, I have seen hungry people in several European cities. If food were distributed more intelligently, it would result in less waste, fewer hungry people, and, ultimately, a lower carbon footprint due to less food being produced unnecessarily,—one of the tenets of the hyper-local food production in Brooklyn. Regional and global socioeconomic inequalities could be ameliorated through better food distribution, which, in turn, could improve human health and the environment.

Recognizing this, various governments across Europe have introduced different schemes to overcome issues related to food waste. In France, legislation implemented in 2016 obliged supermarkets to sign a donation contract with charities and banned them from disposing of or destroying unsold food.[21] In Italy in the same year, the senate approved legislation that aims to reduce a million of the estimated 5.1 million tons of food waste by offering tax breaks to businesses that donate food, including products past their sell-by dates as long as they don't pose a risk to human health.[22]

Regrettably, I was unable to find any comparable legislation in the United Kingdom that would have clear health, social, and environmental benefits. Entrusted to lead in securing the future of those most in need are spirited individuals such as Tristram Stuart, and Jenny Dawson, founder of Rubies in the Rubble[23], who takes unloved fruits and vegetables that would otherwise end up in a landfill, among others.

All of these individuals and the people who work with them undoubtedly are reducing negative environmental impacts and increasing consumer awareness. Reaching a tipping point, these kinds of gastronomic initiatives will become more prevalent, together with more responsible consumption at all levels, instigating much-needed and significant change in the course of the Global North. Each and every decision we make, however small, can influence the outcome.

A MICHELIN STAR IN COPENHAGEN

Located in the trendy and unpretentious district of Frederiksberg in Copenhagen, Relae and its sister restaurant across the street, Manfreds, both offer organic and sustainable food. I met Lisa Lov, a New Zealander chef turned restaurant operations director for Relae, which holds numerous accolades, including a Michelin star[24] and number 39 ranking in the World's 50 Best Restaurants list in 2017,[25] as well as being winner of its Sustainable Restaurant Award in 2015 and 2016.[26]

About 75 percent of Relae's menu is based on vegetables, with the remainder being meat, poultry, and fish. The restaurant uses a nose-to-tail philosophy, bringing in whole animals and not letting any parts go to waste. When they buy a pig, they use some of the parts in one restaurant and the remainder in the other. As Lisa explained, "When we have pork neck and pork shoulder on the menu at Relae, there's likely going to be a rillette (similar to paté), pork skin puffs, and so on at Manfreds."[27]

In order to further display their gastronomic talents and zero waste attitude, Lisa gave me an example at Relae. When duck breast was on the menu as a main course, a side serving was a small pancake stuffed with the braised parts of the neck, wings, and legs. The same applies to fish, in that nothing is left to waste: if Relae is using the fillet and loin parts of the fish, Manfreds will likely use the skins or the bellies, then the scrapings for tartar.[28]

The same assiduous attention to detail applies to the vegetables, which Lisa calls a leaf-to-stem philosophy. If stems aren't used at Relae, they are passed to Manfreds and used there, otherwise they are sent to a tiered waste management system, in which large white buckets are used to collect raw vegetable scraps, as well as eggshells and coffee grounds. Whatever doesn't go into the compost bucket goes into a general food waste bin, which is separated as follows: general, cardboard, plastic, and metal.[29]

More restaurants could do this, maybe not all of the steps at first. But compost is a quick win because the decaying fruit and vegetables are rich in essential elements that can be added to soil for new crops, making the food production/consumption/waste cycle more efficient. Lars Jacobsen, who describes himself as a social gardener, collaborates with Relae through a project called Projekt Offside. Using cargo bikes, Lars's team rides eighteen miles from their farm and collect about four tons of organic kitchen waste each year from the white waste buckets at Relae and Manfreds, which is then used to make compost. When they return, they deliver herbs and greens.[30]

Lars also explained the scalability of the model: "We have found a manageable way to compost the organic waste, and we believe that it is very scalable so we in principle could do it for all restaurants."[31]

Moreover, Lars works with people with mental health issues, such as breakdowns caused by stress. With only one full-time employee on the project, currently it is a labor of love, rooted in the hearts of people and the environment. This is another example of what I had sought to discover on my travels, and with this book in mind.

As I searched for more examples, I discovered a Russian-born Dane, Selina Juul, that the Danish government had credited with reducing food waste by 25 percent in five years.[32] Through her Stop Spild Af Mad (Stop Wasting Food)[33] nonprofit movement in Denmark, Ms. Juul is engaging consumers, businesses, and the government, including Denmark's former prime minister Poul Nyrup Rasmussen, to bring about that important reduction.

Changes I saw in the United States, United Kingdom, and Denmark reflect shifts in attitudes at both consumer and producer/supplier levels: it is becoming less socially acceptable to throw away food. Instead of reveling in their own greatness, individuals in countries with some of the highest per capita income and consumption in the world are taking a long, hard look at the inequities that exist around them and understanding that the excesses are harming everyone not only from an environmental viewpoint, but also an ethical one. Growing awareness that eating excessively and throwing away good food for no justifiable reason when millions of people literally are dying of hunger means that change is happening.

I felt inspired and encouraged that people in the Global North had a common sensitivity, sensibility, vision, and, crucially, courage to explore and implement a range of solutions that are urgently needed. Both individuals and businesses can make a difference through the choices they make.

· 8 ·

Cities of the Future

COPENHAGEN WIND

My ears were burning cold and my knuckles were as white as the paper I'm writing on as I pedaled rapidly along the shoreline in Avedøre, a suburb of Copenhagen, the winter wind punishing me for venturing so far out of the city. I grimaced as I cycled past the Avedøre power plant, an omnivorous machine recently upgraded to be able to burn wood pellets. Its white smoke rose, blending into the pearl sky.

Finally, I managed to see one of the wind turbines through the thick morning mist. It was high tide, which meant that I couldn't cross to its base, so I continued cycling and, after a few hundred yards, found another wind turbine that was on land. I parked my bike and stood directly beneath the massive blades, which cut stoically through the air with relentless strength, causing a whooshing sound every few seconds.

I cycled back to the city center to get a view of Middelgrunden, an offshore wind farm composed of twenty turbines: the first ten are owned by Ørsted energy, and the next ten by a cooperative of stockholders.[1] At launch, Middelgrunden had a total capacity of 40MW.[2] It led the way to other farms in Denmark, such as Horns Rev I in 2002 and Nysted in 2004,[3] and more recently Anholt, with 111 turbines and an overall capacity of 400MW—ten times more than Middlegrunden[4] in about a decade.

Through a number of incentives, including price subsidies and tax breaks, the Danish government offers a financial reward to help increase the country's energy independence while reducing its carbon footprint. Beyond this, it also helps to foster a feeling of civic duty as global citizens.[5] Many Danes I spoke to seemed not only aware of the global issues I was researching, but conveyed a sense of commitment to being part of the solution.

In order to gain a greater appreciation for the numbers and the overall theme of wind power, I met with Martin Hansen, an associate professor in the wind energy department at the Technical University of Denmark, at the Manfreds restaurant I had visited earlier. As a world authority on wind power, he undertakes assessments for governments around the world, including a recent analysis for South Africa. By collecting wind data and combining it with complex mathematical models, scientists and engineers are able to accurately predict the feasibility of a wind farm in a given location: "It's an interesting time, the wind industry has grown exponentially. Subsidies could take it to the next level."[6]

According to Hansen, we have enough wind potential to power the planet, with one of the challenges being the variability of wind power: "There are days where the production exceeds the consumption, but there will always be periods where the wind speed is low and backup power is needed. Therefore, large scale storage of electrical energy to cover approximately three days' consumption is desperately needed to become 100 percent self-sufficient, with, e.g., wind energy, and this is the holy grail in renewable energy that depends on intermittent wind speeds or solar radiation."[7]

Another challenge he made reference to is the unit cost of wind energy, which would need to be lower than coal or gas for it to truly take off: "The price from offshore wind is not there yet and relies on subsidies, and if it does not happen within 5–10 years, then offshore wind may not happen. To bring down this cost, there is an increased demand for larger and larger turbines (larger than 10MW) to limit the number of turbines, and thus the number of costly offshore foundations for a given total power of the offshore farm."

Notwithstanding these challenges, the energy game moves as quickly as the large blades that cut through the sky in Danish wind farms, which have proliferated to the extent that they have contributed to an electricity surplus. On July 9, 2015, an unusually high amount of wind was producing 116 percent of Denmark's needs, peaking at 140 percent by 3 a.m. the following day.[8] Germany and Norway shared 80 percent of the excess electricity, and the remaining amount was sent to Sweden,[9] demonstrating how cross-border collaboration through large-scale integrated and sustainable solutions can point us to a brighter future.

I felt enthused by our conversation, especially by Hansen's insights into smart grid technology that can optimize energy consumption, such as washing machines and electric car battery charging, depending on demand. He also told me that if the price of wind energy does indeed come down, then "the transfer to these technologies will happen and hopefully help [to solve] many of our challenges today, climate change, wars because of oil, unemployment, etcetera."[10]

Just a few weeks after my meeting with Hansen, a dream came true: the Netherlands' main train operator, Nederlandse Spoorwegen (NS) announced that all of its train services, which carry six hundred thousand passengers daily, would run on wind energy. NS had signed a tender with Dutch electricity company Eneco to supply clean energy starting in January 2018 but delivered a year early.[11] Had that train been in operation when I traveled from the Netherlands to Germany, I would have taken it to the border with Germany to celebrate this important step forward for the environment and in the process could have reduced my carbon footprint.

AMSTERDAM BIKES AND LIGHTS

Many cities in Europe seem to be clogged with too many polluting vehicles. This includes Paris,[12] which only a few days before my visit had to go into a state of environmental contingency[13] due to atmospheric pollution similar to Mexico City months earlier, offering free public transportation to get people out of their cars; and London, where inhabitants are forced to inhale deadly exhaust fumes. The main issues stem from particulate matter—fine particles emitted from the exhausts of diesel automobiles, buses, trucks, and other large vehicles, together with a toxic cocktail of gases.

Amsterdam's ambitious plans for creating a zero emission city from 2030 is testimony to a commitment to phase out polluting vehicles in the city,[14] but I don't necessarily believe that electric vehicles need to be part of the solution, due to numerous issues I witnessed with them (chapter 6). Already graced with an integrated transportation system of trains, trams, and buses, bike users can switch from modes of transportation with ease, thus solving the "last mile" problem (how to travel from the final transportation hub to the final destination). The government even offers tax incentives for people to bike to work,[15] which not only helps Amsterdam and other cities to be cleaner, but motivates its citizens to be healthier and shows what can be achieved when education, health, and ethics are considered more important than low-cost convenience.

Standing at an intersection, I saw the movement of automobiles, bikes, and pedestrians moving in a kind of symphony of human displacement. The bikes were ridden by the young and old, women and men alike. Some bikes had extensions at the front or the back to carry more goods. Later that day, I even saw one with a in-built cot for an infant. A few others had trailers behind, making it clear that cyclists not only have confidence in the infrastructure, but in other riders and car drivers.

The Netherlands has an outstanding bike infrastructure. The integrated transportation system makes it easy to cycle to the train station and either leave one's bike there or take it on board, making it a greener,

A view of parked bicycles and a canal in Amsterdam, the Netherlands. Photograph by author, 2016

and sometimes cheaper and quicker, way of moving around compared to using buses or automobiles. The dedicated bike lanes on either sidewalks or the road offer cyclists more protection and freedom than their counterparts in other cities around the world and are commonplace across Amsterdam.

In both Amsterdam and Copenhagen, it seemed intuitive to establish a link between the innumerable bikes I saw, high educational attainment, a great sense of civic duty (for example, most streets I walked on were litter free), and the first-class public transportation system, with clean air. I then concluded that if children are born in this environment, where bikes are more important than cars and where parents literally are leading by example, that the next generation will be equally, if not more, connected with the environment.

Berlin, where I currently live, has a good bike infrastructure, and it's perfectly normal to go shopping or to meet a date using a bike. Still, it seems to have relatively more cars for the obvious reason that some of the biggest and most prestigious carmakers in the world are headquartered in Germany and provide a lot of employment, helping to ensure that the country is one of the richest in the world. These factors perpetuate the duality of values that exist in the country today: green/hydrocarbon, liberal/right wing, fair trade/exploitative trade and foreign policy, organic/GMO, bespoke/mass production, the haves/have nots, which is something that is mirrored both in the United Kingdom and the United States.

An increasingly familiar sight, be it in Mexico City, Atlanta, New York, Berlin, and other cities around the world, are public bike-share schemes whose aim is to provide a convenient and viable alternative to using fossil fuel-dependent, hybrid, and electric vehicles, all of which have numerous downsides (chapter 6). Such schemes have numerous benefits. For example, a recent study in Barcelona showed that the city's bike-share scheme reduced twelve deaths as a result of the cyclists' physical activity and saved more than nine metric tons of CO_2.[16] I used a similar system in Mexico City by purchasing a card that enabled me to pick up any bike and use it for up to thirty minutes at a time free of charge.

A variation of the systems that use fixed stations are dockless bikes such as Mobike,[17] which can be parked anywhere within a designated zone in numerous cities around the world. I use Mobike, which allows me to jump on any available bike for thirty minutes. The scheme's app tells me how many kilometers I've cycled, calories I've burned, and the CO_2 I've saved on each trip, as well as a grand total for each of these figures. Cell phone and application technologies such as this have made strides forward and converged, enabling phone apps to do these kinds of calculations, something unimaginable a decade ago.

But, as a good friend pointed out to me, these dockless bike systems are provided by startups that offer artificially low prices, which can squeeze out local bike shops that sell, rent, and fix bikes. This is the dark side of globalization. Perhaps a middle ground is possible where bikes are built and maintained locally, decentralizing the capital flows and ensuring that everyone gets a piece of the pie, which is something I have yet to see.

Moreover, the bike utopia I witnessed in Amsterdam, Copenhagen, and later in Berlin, is not true in other nearby European cities, such as Paris and London, where the effects of cars are all too apparent. Road deaths are a concern for cyclists partly due to lack of infrastructure, awareness, and respect for them. The situation was similar in Mexico City—whenever I rode a bike there, I was taking a big risk. In response, more and more cyclists attached cameras to their crash helmets and wore high-visibility jackets to help create more awareness.

It's clear to see that the number of bikes, which for shorter journeys offer similar levels of freedom, if not more, than the car, are sparse in comparison to their insanely polluting analogs. If we could get people out of cars and into bikes, the tangible benefits, including reduced morbidity and illness, through increased exercise and improved air quality would help society. The solution to personal transportation in big cities is before our very eyes. The next step is for more people to swap gas pedals for bike pedals.

While taking photos of bikes around Amsterdam, from cargo bikes to tandems, and other types that I had never seen before, I accidentally found myself in the red-light district. It was midday on a typical gray winter's day. Driven by curiosity, I looked into one window and saw a woman with broad shoulders, large hands, and her naked body. Her bolt-on breast implants. Her tan. Her empty eyes. Moral judgments aside, I wondered what had happened to consumerism.

A sign outside an apartment building a few doors down from a brothel read: "Please don't piss/puke here, we live here." Were the drunk or stoned young men I saw walking on the other side of the canal in any way interested in women's welfare? Was the woman I saw happy doing this job? Was she there of her own free will? As unpalatable as it might sound, I asked myself why I was unaware of caged men offered for sexual rent to drunk or stoned women.

As I wandered along Amsterdam's beautiful streets and crossed its picturesque bridges, I continued to reflect on the prostitutes, who reminded me of the ubiquitous vending machines around the city. The human body has become just an entertainment service similar to Netflix, or a fairground ride that can be paid for by the consumer and taxed by the state. This was another side of Europe, beyond the Van Gogh galleries, Michelin-starred restaurants, pristine parks, and palaces where kings and queens still live. Human trafficking and other forms of human exploitation are an almost silent, invisible part of that complex landscape.

One reason why this form of "sex and drugs" tourism is popular is because air travel in Europe is so cheap. Ultra-low air fares are possible through an intricate web that includes bank loans with historically low interest rates, aircraft manufacture, materials (chapter 14), global supply chains, marketing, and cheap oil. Because the environment is not taken into account, demand has exploded; air travel is expected to double in the next fifteen years. Ryanair, a low-cost European airline carrier, is now on the list of Europe's top ten biggest polluters,[18] as Antarctica continues to melt at a record pace.[19] As mentioned in chapter 6, the airline industry was susceptible to the initial economic effects of the coronavirus, meaning that these forecasts will need to be recalculated once any semblance of clarity emerges in a post-coronavirus world.

The second reason why such tourism is possible is that the state decided sex workers should be taxed. It honestly felt like I was walking through a zoo. I pitied the workers, not just because they were semi-naked, but also because of the emptiness in their eyes. I could see a strange sort of sadness in them. The windows with neon lights were cages, and the pimps, nowhere to be seen, were the zookeepers; the state had become the revenue collector, immune and desensitized, collecting its taxes. What had freedom, choice, and dignity turned into?

Another reason why tourists can flock to this haven of senselessness is that drug consumption, whether illicit or otherwise, is very popular in Europe. Street cocaine quality has increased,[20] as has consumption.[21] While both wholesale and retail cocaine prices between 2007 and 2017 across the continent were fairly stable, differences can be seen between European countries: cocaine in the Netherlands is nearly half the price it is in the United Kingdom.[22]

All of this has come at the detriment of the environment, especially in South America where the white powder is sourced, not to mention the gruesome drugs industry and its inhumane violence inflicted on fellow human beings, where absolutely no mercy is shown (chapter 5). But the consumers don't see this. They're way too high and don't care about anything other than their own entertainment.

Despite my unexpected detour in this less-agreeable part of the city, I left Amsterdam feeling uplifted by what I had seen. The locals, who seemed committed to the environment by not using cars and raising the bar even higher, enabled me and others to see that it is possible to live in a modern city and not rely on cars. The optimism I harvested would provide me with useful inertia to propel me across Europe, where I would have some difficult conversations about other ways in which people are forced to move in our world, before reaching Istanbul, the final destination of this leg of my journey.

· 9 ·

Crossing Europe

CALAIS CAMP AND REFUGEES

*A*fter carrying out interviews in London related to zero waste (chapter 7), I initially had planned to travel to Calais, France, to interview refugees in the "Jungle," an infamous refugee camp located next to the port.[1] A few days before buying my ticket to France, the camp was forcibly shut down by French authorities, displacing people who had come from Syria, Iraq, Afghanistan, and elsewhere who were looking for a better life in the United Kingdom.

I spoke with Roz McGregor, an international relations master's degree student at the London School of Economics, who had taught English at the camp just a few months before its closure. Using her own resources, Roz traveled to France to provide support through teaching English to mostly Middle Eastern males and found the conditions challenging: "In a way I'm glad that it was closed down before winter . . . because I can't even imagine, actually, what it would have been like to live there in winter."[2]

Due to the lack of sanitation, the late-summer heat combined with the overcrowding meant that the smell was awful. The camp was muddy and dirty, but in spite of this, Roz took herself out of her comfort zone and dedicated her time to those who had to flee war and persecution. I asked her whether the experience had changed her. Perhaps she played it down by saying that it hadn't per se, but I suspected it actually had. While sitting in an empty restaurant on the ferry back to the United Kingdom, she expressed to me how bizarre it felt that she was able to be there just by showing her passport, while other people, just a few miles away, were dying trying to get on the seat she sat on.[3]

I, too, share this privilege, as do many other passport holders in the Global North. We are able to cross borders easily, sometimes under a visa waiver program, other times by receiving an automatic visa on arrival, only in a few rare cases perhaps having to visit an embassy and complete the necessary paperwork before arriving at the country of our choice. Before departing, I asked Roz what she thought might be a simple practical solution to change people's misconstrued opinions about refugees. This was her reply:

> I know there's no way the government could do this, but if everyone had to go and spend even a week at the refugee camp it would transform, I would imagine, people's thinking because—I know this is a bit of a cliché—it's much easier to hate or resent people when they're faceless. Once you actually get to have a conversation with someone and actually understand why they're doing what they're doing, you realize that most humans would do what these people are doing.[4]

Roz added that although there may be logistical challenges related to letting every refugee into the United Kingdom, as global citizens we need to work collectively to see how these vulnerable people can be looked after—that it is a moral responsibility to do so, and that empathy has to be at the heart of the overall solution.[5]

Today, we have the greatest number of refugees and displaced people in the history of humankind. In 2019, the number of refugees worldwide stood at twenty-six million,[6] equal to three times the population of Switzerland, bringing the total number of displaced people that year to 79.5 million.[7] War, human rights abuses including torture and discrimination, environmental issues, and extreme poverty have given an increasing number of people no other choice but to leave.

As modernity has brought with it fast cars, planes, and trains to transport us rapidly across continents and beyond for business or pleasure, so, too, have the means by which people can escape increased. Boats frequently sink crossing the Mediterranean, be it the North African route to Spain or Italy for African migrants, or those filled with Middle Eastern migrants in the waters around Greece.

Not only are the figures related to refugees and displaced people a cause for concern, as well as the challenges they face both at home and in their host country, but it is the acceptance of this state of affairs that is unnerving. During the 2016 Olympic games in Rio de Janeiro, ten refugees from South Sudan, Ethiopia, Syria, and other countries were allowed to participate in the games in a newly formed team called the Refugee Olympic Team.[89] We now watch refugees perform as teams and cheer them on thinking that we are being empathetic and showing solidarity, but in doing so, have we become complicit in normalizing their predicament.

Consider the strangeness of the situation. Some 165,000 Sahrawis are stuck in the desert in the Tindouf refugee camps in the south of Algeria[10]—where Al Qaeda are active—waiting for an outcome for the Sahara State for more than forty years. Palestinians stuck in Shatilla refugee camp in Lebanon and other camps around the Middle East for more than seventy years, Syrians in camps across the Middle East, and ethnic Rohingya Myanmar people in Bangladeshi camps are just a few more examples.

Humanity is witnessing new kinds of distortions, which have somehow been digested and accepted as being part of the twenty-first-century landscape. The temporary has become permanent, the transitory has mutated into perpetuity, and the perverse has become silently accepted. I wonder if this trend will continue, and whether one day a drop-down box titled "Refugee" as nationality will be a choice in an online transaction. Perhaps we'll have a refugee passport as well? Will we reach a point in humankind when the number of refugees might be greater than thirty million, fifty million, a hundred million? At what point do lives, or numbers, matter?

I met with Saba, not her real name,[11] a thirty-four-year-old Syrian mother of three, at her home in a middle-class neighborhood in Bratislava. I wanted to avoid the painful clichéd questions related to life in Aleppo, which reached horrific depths of human depravity in the months that preceded her departure. With her two daughters sitting close by in their open plan apartment, I wanted to learn about her journey as a single mother who entered Slovakia as a refugee, rather than focus on the traumas of her hometown, which is widely known and documented as being one of, if not the most, vicious cities on the planet.

With typical Middle Eastern hospitality, Saba brought out Turkish coffee, water, and some biscuits and, holding back tears, began to talk about her journey. She left Aleppo overland to Turkey with her daughters in March 2016 and made it to Istanbul, where they spent ten days. They then took a boat to Lesbos, one of the Greek islands closest to Turkey, where they spent three nights, then traveled overnight by boat to Athens.[12]

After finally starting the process to gain refugee status there, which took two months, they traveled to another Greek island, Rhodes, then returned for an interview in Athens and eventually decided to take a chance on Slovakia. "It's not our homeland. Turkey was similar to Syria, Greece a little less. Slovakia is very far away from where I'm from."[13] The ice-cold December winds, lack of sunlight, and a completely different way of living were all alien to her.

For Saba, the biggest challenge to integration was language. As I walked through the front door, I could see pink, green, and yellow Post-it notes across the kitchen and reception area. Words in Slovak, such as *topanka* (shoe) and *pohar* (glass), had the translations written beneath them in Arabic. Saba laughed with her eldest daughter as she struggled to say the "p" in those words, because this letter doesn't exist in Arabic.

Her ex-husband had left Syria three years earlier and headed to Germany, where he was joined two years later by their son, whom she visited a few months ago. Through video calls, Saba keeps in touch regularly both with her son and family members who are still in Aleppo. In spite of the distance from both Syria and her son, the linguistic challenges, financial constraints, being homesick, and the stigma of being a refugee, I was surprised that Saba felt optimistic about the future. Perhaps the realization that her children are safe in Europe means that an immediate existential fear has been placated. Beyond this, she has received support from the local community and has had the chance to integrate, slowly making friends. Before saying good-bye, I asked if she hoped to return to Syria one day. "Inshallah," she said. "God willing."[14]

Later, I spoke with Barhum Nakhlé, who helped me with my Arabic during the interview.[15] He works as a translator with Saba and other refugees in Bratislava. Although his official responsibility is as translator at the migration office of the Republic of Slovakia, he also does social work. I asked Barhum whether Saba's largely positive experience was typical for a refugee. He didn't seem to think so. A Somali woman had suffered physical abuse on the public transportation system because of her skin color,[16] which immediately reminded me of my visit to the Martin Luther King Jr. Memorial Center in Atlanta during a day off from work, and the freedom riders such as Rosa Parks who went against the grain and rode the public bus to protest segregation. Of course, the story in Slovakia is different, but the essence is the same: hatred and violence.

A Slovak national himself, Barhum explained that things would be very different for Saba had she been veiled and/or come accompanied by a Syrian man; and seeking asylum, in his view, is easier for Christians. That she was a single mother and an unveiled Muslim woman were key factors to her being allowed to stay in Slovakia.[17] As I departed, I exchanged a warm hug with Barhum, but I had mixed feelings as I traveled through the night toward Serbia. I had spent Christmas with Barhum and his family, a time to reflect on the Christian tradition of sharing and being together, and considered how lucky I was compared with those who didn't have a passport, a roof over their head, clean water, food in their belly, any semblance of safety, or someone to hug them.

LIQUID FEAR

We face a grave peril with regard to how we view, process, and respond to the stories of refugees. Today, the headlines on both sides of the Atlantic are about illegal immigration, terrorism, and how bad and dangerous the "other" is. The very same bullets that were made in Europe and the United States, and are regularly fired by them, either killing or driving people out of their homes, are increasing the numbers of refugees around the world. Moreover, the relationships that the United States and Europe have with governments such as Saudi Arabia and the United Arab Emirates (chapter 15), which have become major military players in the region and are key buyers of weaponry,[18,19] exacerbate this phenomenon.

Nothing is worse than hypocrisy, which is why I cringed when I saw Prince Charles, a staunch environmental advocate, join a sword-dancing ceremony in Saudi Arabia[20] while knowing that the country is responsible for numerous human rights violations such as beheadings, lack of women's rights, and restrictions on freedom of expression, among others, which would be completely unacceptable in the United

Kingdom or any other so-called modern and egalitarian society. And there he was, looking like he was having a whale of a time while people were suffering or ceasing to exist.

Further, consider the systematic theft of oil in countries that have collapsed, including Libya, Syria, and Iraq, some of which ends up in automobiles, buses, airplanes, and heavy industries in foreign lands. It would, therefore, seem intuitive to suggest that if consumers knew that about these other elements in the production of oil, and understood the geopolitical effects of oil (including and beyond the Middle East), perhaps they would take steps toward reducing their use of gasoline. Maybe they would walk to the grocery store instead of driving. Or, perhaps, the idea of making informed choices is naive and sentimental. Either way, today's consumers have been cut out of the loop completely.

A recent example of callousness, based on the principle of excessive and unwarranted use of violence against a relatively weak adversary, was the invention of the "axis of evil." Through it a fictitious enemy was born. Saddam Hussein didn't have weapons of mass destruction that could be used against the United States or Europe. Not only did the war with Iraq, which began in 2003, cost the American taxpayer more than US$2 trillion,[21] which could have been used to support the millions of Americans living in poverty, not to mention other parts of society that could have benefited from more affordable health care and education. But, as touched on earlier, families in the United States, United Kingdom, Spain, Italy, Denmark, and elsewhere lost soldiers. They were coaxed into thinking that they were protecting their nation and democracy, but they were doing anything but that: it's an open secret that the war in Iraq was about stimulating these economies through military spending and ensuring the global supply of oil, which help reduce inflation.

Governments continue to find bogeymen as pretexts to invade privacy, as well as use excessive force at home and abroad to help protect their people. Yet, strangely, other forms of extreme violence, such as the invasion of Iraq, have led to no criminal convictions. That outcome is in spite of the lack of legality of aggression toward a sovereign state and the hundreds of thousands of dead Iraqis, not to mention the thousands of dead servicemen and servicewomen duped into fighting a war that was completely baseless and unnecessary and, therefore, immoral. Let me expound further: if the defense industry is, effectively, the apparatus used to decimate human rights in the Middle East, with European and

United States governments fighting unnecessary wars and doing business with nations such as Bahrain and Saudi Arabia (who sell cheap oil in return for military equipment and weapons to crush opposition both at home and abroad, such as the Saudi led war in Yemen), then this creates a paradox in European morality, which is supposedly based on the Christian and human values of justice and equality.

At an individual level, we watch the death counts on the news as if they are football scores and react depending on which side has lost. ISIS versus Iraq is different than ISIS versus France. Our anger isn't dependent on the final death toll but, rather, on how it affects our lives. Could I have been in that spot? Did the people who were killed belong to the same tribe as me? If the answer to either question is yes, then it's an issue. If the answer is no, then it's far away and, therefore, someone else's problem.

Perhaps this illustrates how Roz's faceless refugee would come to light if we could meet people and learn from them. Ignorance, apathy, fear, and disdain help to maintain this state of affairs. Nothing will ever change if the roots of these problems aren't pulled up from their depths and replaced with education, empathy, and commitment. The refugee crisis is a symptom, not a disease in and of itself. Once the bombs stop dropping across the Middle East, it's highly likely that some semblance of order would be restored. People could return home and futures could be built.

As the world becomes more intertwined through families that span countries, time zones, continents, ethnicities, and religions, the interconnectedness of our individual experiences will affect our collective existence. This, in turn, introduces an unimaginable peril as climate change catches up with other people. It leaves them with no other choice but to flee their homes because they are no longer sustainable for living in, such as those who reside in El Hatillo, next to the open-cut coal mine in northern Colombia (chapter 3).

This brings me to my final point on the subject, which is Europe's consciousness today. A massive fissure has created a clear distinction between ecologically oriented behaviors, in which people are concerned about the environment and other global issues such as human, animal, and environmental rights, and actions such as obtaining goods at the lowest possible price without regard for the environment or human rights outside the European Union.

Naturally, a large group of individuals are scared to interact and make a real change. They suffer because they know, deep down, that systematic failures and injustices are happening, and they have become paralyzed, believing that they can do nothing. In this sense, migration, war, pillage, and the climate crisis have been successfully manipulated to create a smokescreen that allows people to avoid talking about the single most important thing that needs healing: our relationship with the environment.

———— ∞∞∞ ————

As I traversed the eastern European countries of Slovakia, Hungary, Serbia, and Bulgaria, aiming to reach Istanbul before New Year's Eve, I could sense quiet in the air. People were rushing home to spend time with loved ones. I didn't have much to do besides buying food and water and organizing my accommodation and transportation, which in some ways was a pleasant departure from the trip thus far given that I had to overcome language barriers even to find a cheap hotel near the main station in Sofia, Bulgaria, at midnight during heavy snowfall. The triviality of my problem actually enlivened and enriched my spirit.

The empty streets of Belgrade and Sofia were very different from Copenhagen and Amsterdam. More automobiles seemed to be polluting more, as evidenced by the smoke coming from the tailpipes, and seemed to be a lot older. I didn't see organic shops, shared bike schemes, or any of the other things that I took for granted such as the abundance of pristine parks and well-manicured sidewalks. As I walked around both cities, I reflected on how Europe was changing with respect to time. The continent was headed in different directions: modernity/tradition, free markets/farmers' markets, wealth/poverty, I/us, war/peace, xenophobia/tolerance. And there I was in Eastern Europe, frankly wondering exactly how many people were part of Europe's plan for a robust continent in which freedom and economic prosperity reigned, and how many had been left behind, forgotten.

My sense was that there were more of the latter, who represented the underbelly of an iceberg of discontent, lack of opportunity, and stagnation, creating fertile ground in which feelings of isolation and resentment could flourish and release a poisonous, contagious fragrance into the air, causing a fever of hate. Then I recalled that as humans, we love stories. What better story for a child than a simple, unforgettable one such as Goldilocks and the Three Bears, Little Red Riding Hood, or Jack and the Beanstalk. For adults, the recipe remains unchanged in terms of the stories sold to us by the mainstream media. By keeping the number of characters low (us and them), and the story simple (they are going to steal from/rape/annihilate us), the masses accept the fairy tale as fact and completely miss a real problem that they and every living thing on the planet face: the sixth extinction.

Populist votes for right-wing politicians in Europe and the United States in recent years reflect an entrenched fear of the "other." This, together with unfavorable economic conditions, helped secure the Brexit victory for Nigel Farage[22] and ultimately Boris Johnson (United Kingdom), the presidency of Donald Trump (United States), the rise of other figures such as Marie le Pen (France), Geert Wilders (the Netherlands), and other charismatic and benevolent "saviors." Their mandate was to help rid society of the perceived problems: too many nonindigenous (i.e., nonwhite and or non-Christian) citizens changing the cultural values of their host country and taking their jobs; Islamic fundamentalists carrying out terrorism; refugees placing a burden on societies, including their supposed inability and unwillingness to integrate.

The convenient and reductive narrative relates how the "dangerous" foreigners will come to sell drugs, rape, kill, and/or carry out a terrorist attack. Although it is true that these things have happened at the hands of foreigners, they do not reflect a real bias toward crimes they carry out compared with crimes by locally born people. In fact, a study in England and Wales in the 2000s showed no link between crime and immigration.[23] By playing on people's fears of too many dangerous, ungrateful, and lazy individuals who have brown skin and may be Muslim—and, therefore, present an existential threat—politicians, with the support of right-wing media outlets, have helped put brown people, especially Arabs, and to a slightly lesser extent, Latinos, at the center of the canvas of their worst fears. Sociologist Zygmunt Bauman described these concerns as liquid fear:[24] an exaggerated preoccupation that something bad can happen anywhere at any time, such as terrorism.

Sadly, the perceived threats related to immigration and terrorism are disproportionate and have led people navigating difficult economic times—including high unemployment rates in G7 economies, high living costs, and low disposable incomes—to view migrants, many of whom have made positive contributions, through hostile eyes.

Economic migration has underpinned the development of many countries: the United States has benefited from an endless wave of migrants from all over the world who want a better life and are willing to work long hours, and in many cases do the jobs that a local person would be unwilling to do, such as domestic cleaning and shift work. At the other end of the spectrum are the dozens of American Nobel Prize winners born on foreign soil, not to mention all of the scientists and engineers who work in Silicon Valley and at NASA who hail from the four corners of the planet, and others who have helped the country excel in science, technology, medicine, and the arts, among other fields.

The United Kingdom, too, has had numerous waves of immigrants: Caribbeans in the 1950s, Pakistanis and Indians in the 1960s and 1970s, and Poles in the 2000s, the latter causing ill feelings among Brits who felt that they were taking their jobs once Poland joined the European Union in 2004. Other countries across Europe followed similar patterns of cheap labor: Switzerland hosted a wave of Spaniards in the 1970s, Germany received Turks in the 1970s—the list goes on. Not only did these economic migrants help reduce inflationary pressures due to their comparatively low wages, but they also contributed culturally, and they have helped to increase the levels of heterogeneity in their new homes.

With that in mind, and considering the factors that truly do threaten these modern societies, terrorism statistics in Europe pale in comparison with other countries, particularly the Middle East and certain parts of Africa, which are home to the greatest numbers of refugees. Germany, which hosts a significant number of refugees, is the only country in the European Union in the top ten.[25] The terror problem is acutely problematic in Nigeria and Iraq, and yet we have chosen to stay silent in the face of these horrors and continue with business as usual: importing oil at low prices and exporting goods, including weapons, to governments with appalling human rights records.

Meanwhile, war is being waged against the environment by those in power and those wanting to be in power. Climate denial has been an important tool in the anti-environmental lobby of Nigel Farage, Donald Trump, and other political party leaders who have preferred to distract from the single most important existential threat since the dawn of human civilization to oversimplified discourse drenched in lies and hatred. The tragedy also lies in the outcome that these discourses have successfully stirred people's emotions and reinforced the binary "us and them" narrative to such great effect.

END OF THE LINE

I finally reached Istanbul a few days before New Year's Eve. The city had modernized even further since my visit four years earlier. The metro system had improved significantly with new stations, as well as an expanded metrobus system similar to those in Bogotá and Mexico City, which have dedicated lanes criss-crossing the city, making it a quicker and cheaper alternative to the car. However, I was concerned by the thick, black smoke that the double-length buses were emitting, given the revelations during my trip across western Europe. The halo of subsequent pollution above Istanbul, similar to other heavily polluted cities such as Los Angeles, São Paulo, Mexico City, and others, was yet another familiar sight.

The atmosphere in Istanbul was tense following the numerous threats that the country had faced in recent years.[26] As I explored the city, I found that several businesses along the famous Istikalal (Independence) street had shut due to the economic slowdown, due in part to the negative press caused by terrorism. A week later, I woke up to the terrible news that thirty-nine people had been killed at the Reina nightclub.[27] It emerged that the attacker allegedly chose the exclusive nightclub at random after being unable to strike at Taksim Square, the equivalent of Leicester Square in London or Times Square in New York, due to heavy security.

After crossing the Bosphorus, the stretch of water that separates the European and Asian sides of the city, and the only route through which large ships and oil tankers can access the Black Sea from the Mediterranean, I reflected on the country's energy plans. Although Turkey derives most of its energy from nonrenewable sources, it draws on its geographical features and produces about a quarter of its energy from hydroelectric plants.[28] However, even that is a risky strategy because rainfall is unpredictable and, as I would discover in Zambia (chapter 13), overdependence on hydroelectric power, or any single renewable source for that matter, can have a major downside when weather conditions change unfavorably.

While I was walking around neighborhoods on the European and Asian sides of Istanbul, I realized how similar Istanbul was to Buenos Aires and other Latin America cities owing to its array of museums, art galleries, and shopping malls, albeit with its own distinctive flavor and style. Perhaps the most striking parallel was lots of young people. They were wearing, to a greater or lesser extent, fashionable clothes, using the latest mobile phones connected to super-fast mobile internet, and represented popular culture and modernity. I wondered what tomorrow would bring for all of them given that global populations, carbon emissions, and temperatures all were increasing.

Each country I visited in Europe has geographical strengths and weaknesses. Denmark might not be blessed with as much solar potential as other countries in Europe, but it has exploited its endowment to become a net exporter of wind energy on those days when it has a surplus and shown its leadership on the world stage of renewable energy. Nevertheless, the onus shouldn't be on any single country to change, as is the case with Denmark and the Netherlands, which are light years ahead. But all countries should contribute to reducing fossil fuel use because it kills people due to the environmental hazards it creates and/ or the violence and crushing authoritarianism by its custodians, not to mention the pollution it creates once the fuel is burned. We are all affected in the long run.

The European project that helped to rebuild Europe financially, and develop it socially and politically, after World War II is now under siege. This project began with so much promise. Success and defeat in World War II were put in the past. No checks at borders within the Schengen area, the cheap flights across Europe with Ryanair, Easyjet, and other low-cost carriers, the abolishment of extortionate mobile roaming fees for inbound and outbound calls, and the continuation of the Eurovision song contest, among other initiatives, fostered feelings of togetherness, as well as a willingness to accept the comparatively small differences between languages and cultural leanings. Historical enmity and animosity thus were swapped for trade alliances, banking, Airbus, the Eurofighter projects, the Channel Tunnel, the wars in Iraq, Syria, Afghanistan, and Libya, all demonstrating a broad range of skills from the art of engineering to war, illustrating that the peoples of Europe could join forces and together create something epic.

But this has long had a dark side. Europe, the kind and benevolent bloc that it is, turned a blind eye to what happened in the former Yugoslavia in the 1990s due to the simple fact that Muslims were being killed. The pink washing by supporting some gay rights, and the green washing by supporting some environmental rights, are entirely consistent with Europe's hypocrisy, both past and present. When the *Concordia* ran aground in Italy, the victims received all kinds of support, and Captain Francesco Schettino, who jumped ship and left innocent people behind, was convicted.[29] In stark contrast, German Captain Pia Klimp provided safe passage across the notoriously dangerous Mediterranean between Libya and Italy to men, women, and children who had to flee unimaginable human suffering. The *Iuventa*, the boat she was captaining, was seized in Sicily in August 2017. In addition, Captain Klimp faces trial and is at risk of a significant fine and possible imprisonment of up to twenty years for her humanitarian deeds.[30] Meanwhile, Europe wants to further bolster its borders. Why? Because the passengers in question are African and Middle Eastern. Undesirables.

Then Brexit happened as I was traversing the continent. Nigel Farage, himself married to a German,[31] and seemingly shy to speak French even though he lived and worked in Belgium as a member of the European Union, wanted to Make Britain Great Again. Through his lies and cunning, like the Artful Dodger, a Dickensian protagonist, he achieved the unthinkable: he was able to free the United Kingdom of Great Britain and Northern Ireland from the grips of Europe and emancipate his people, delivering them from evil in the form of migrants, like a God. Ironically, only a few years later the coronavirus would make migrants, who are part of the DNA of Britain's beloved National Health Service (NHS), all the more valuable, as I would discover toward the end of my journey (chapter 16).

The United Kingdom, already making huge strides forward in its Brexit procession full of pomp and circumstance, declared that it would be carbon neutral by 2050.[32] Such a languid approach to a mass extinction event that has already begun reflected how out of touch the ruling elite of the country were with basic science. At least the behavior was predictable and consistent.

Other EU countries demonstrated a variety of attitudes: Germany announced that it would stop extracting and burning coal by 2038,[33] while others such as Poland said it wouldn't be able to even meet the EU's 2050 target and would continue to develop its coal industry, which provides around 80 percent of the country's energy requirements.[34] In any case, the 2050 zero emissions target would be like scheduling a meeting in 2050 with neighbors who live in an apartment complex that is ablaze in 2020: it all sounds very promising, but let's not fool ourselves; the only interests are to increase fuel extraction in order to stimulate the economy and get reelected the following year. Voters aren't interested in the details but want to have a good quality of life, and rightly so. The only problem is that the notion of this quality is unsustainable as it currently stands, and nobody has yet explained that to them clearly.

Meanwhile, Norway, which is not part of the EU, proudly announced that its new refinery, the third biggest in the country, would operate using renewable resources: another example of a smokescreen to give the impression that progress is being made when, in fact, governments are just pulling the wool over the public's eyes.[35] If Norway, given its high standing in the international community and the financial resources it commands through its sovereign wealth fund, which stands at more than one trillion dollars,[36] would divest from oil, this would send positive signals to Europe and the rest of the world that there is another way forward. However, Norway having temporary wealth from oil has proved to be more important than making a serious investment in renewable energy that could both safeguard Norway's financial future and increase the chances of there being a livable planet for everyone in one hundred years. This was another missed opportunity.

In an attempt to catch up with China, the United States, the world's second biggest carbon emitter, began to relax environmental laws during the Trump administration, including: revoking California's ability to limit exhaust emissions on certain vehicles;[37] revoking an Obama-era federal sustainability plan, which would have reduced the federal government's greenhouse gas emissions by 40 percent over ten years;[38] and reversing an Obama-era plan to reduce methane emissions,[39] along with a long list of other decisions that can be seen as detrimental to many Americans, the rest of humankind, and the planet.

The most forthright of all of the United States' changes was to pull out of the Paris Climate Agreement,[40] thereby unshackling and freeing its economy so that it could fulfill the promise of its limitless potential. Unabated by any ceilings or worldly constraints defined by the laws of nature or simple reason, American industries would have carte blanche to pollute as they wished once again. Many Trump voters were sufficiently distracted with discourse related to rapist Mexicans[41] and fanatical Arab terrorists, both impediments to making America great again, together with support from Russian hackers backed by the Kremlin who derailed Hillary Clinton's presidential campaign and manipulated data on social media, among other stealthy and insidious actions. Through this, voters came to believe in nonexistent risks born from hate and lies, reflecting their anxieties at the ballot box. This is democracy in the twenty-first century.

In light of all of this, it didn't come as a surprise to me when, living in Berlin a few years later, I saw a "Returning from Germany" poster campaign by the German government[42] plastered across the city, which offered a financial reward to lost souls from Syria, Afghanistan, Pakistan, and other countries who would agree to go back home. What home? Nor did it come as a surprise that I was threatened with a knife by a skinhead in Reinickendorf, a quiet neighborhood with a large migrant population in Berlin, in broad daylight.

And it definitely came as no surprise when I learned how many American and European troops were still active in Libya, Syria, and Iraq, all trying to get their hands on the black gold that helps to support their dirty carbon economy and privileged way of life. The second Gulf war told them that no force was strong enough to stop them. I am sorry for the hundreds of thousands of people who died, the millions who were displaced, and the thousands of troops who were duped into fighting in a wholly illegal and unholy war.

What really astounded me during my time traveling across the United States and Europe was the degree to which people were commodified. I discovered that prostitutes could be ordered online as if they were pizzas—pick a size, color, preference, price, and they'll even come to your home. The hungry and homeless, from Atlanta to London, who don't have a bright future are now a permanent fixture on the canvas of any modern city in the Global North. The shift workers I saw in the bus and train stations were almost always nonwhite. The desperation in some of their eyes reminded me of the damaged environment I saw in Latin America.

My only hope was—and still is—the wonderful hearts and minds that I had met along the way. They filled my soul with hope. Luckily for me, I would meet more inspiring people later, even if those meetings would sometimes be unexpected or, in one case, delayed by three years. The path is never straight, clear, or certain. I always try to be vigilant and careful, taking notes and making observations about what's going on around me, paying particular attention to the good things, the brave people, and those who love and care, for it would be these people who would stay in my heart throughout my journey, and who I hope will remain there forever.

CITY OF FLOWERS

About a month before I had planned to travel to the United States for the second leg of my world trip, I began to feel chest pains when I was back at my old home in Mexico City. I visited a physician in a nearby hospital who began a series of tests over the course of a couple of weeks and concluded that I had severe erosions in my esophagus caused by acid reflux (heartburn). I scheduled the recommended surgery to take place in four months. This severely compressed the next leg of my journey, but with relatively good transport, clean food and water, as well as friends I have in the United States and Europe who would end up hosting me, I decided to roll the dice and went for it. During that trip I ate more than usual, fattening myself up in the knowledge that I would shed the weight soon after my return.

The months after the operation were frustrating. I loved being on the road, but being out of commission and having lost more weight than expected because eating was so painful and my stomach had shrunk meant that I would need about eighteen months for it to get back to its original size. Rather than being angry and dwelling on what I couldn't do, I tried to use my time fruitfully and took a break from reading papers about climate science and demographic forecasts in different countries and regions, reverting to one of my heroes: Erich Fromm, existentialist extraordinaire. After rereading *The Art of Loving*[43] and *To Have or To Be*,[44] I reflected on my life. What did I truly love? What was precious to me? Why did I get sick? I had a mechanical problem with a valve in my esophagus, but that wasn't the only thing. It was somatization: the Earth was sick, and so was I.

The first few years of research for this book, which were mainly focused on global levels of carbon, and fieldwork in Latin America made me worry excessively about the planet's health. So, when I returned from my journey across the continent, I became melancholic and refused to leave my apartment for a month. I didn't call my friends to let them know that I was back. In my solitude I reflected on my journeys, where I had seen heaven and hell, and slowly processed what I had documented.

While editing the photos of the beautiful seedlings in Salgado's forest (chapter 3), I realised that while I was concerned about nature, I was too anthropocentric—and therefore reductive—in that I was only seeing the world through the eyes of humans and putting us at the center of the canvas when, in fact, we are a small brushstroke on the most exquisite masterpiece of all: life on Earth, with all of its colors, sounds, textures, and variations. Buoyed by this realization, during my period of self-isolation, I planted tomato seeds in a couple of flower pots a neighbor gave me as a gift. I checked in on them every day and saw them germinate and grow quickly. I went to the local garden center to buy more soil, then gathered ten-liter water bottles from my neighbors, cutting holes at the bottom for drainage and discovering a new passion. Inspired by the tradition I had learned about in the Mexican jungle, I created my own *milpa* on my roof terrace by planting corn, beans, and zucchini, which work together through a symbiotic relationship and grow very quickly. Gaining confidence and experience, I planted cotton and lots of herbs, such as mint, thyme, and parsley.

As my physical and emotional strength returned, it seemed natural to begin to reach out to other urban gardeners to get tips as I wanted to connect with other people and nature. These conversations slowly transformed into a series of about fifty interviews that I conducted in Mexico City. With each interview, my depleted reserves of optimism were replenished, such as when I spoke with Elia Martinez, a gynecologist who quit her job in a Mexican hospital and decided to join Doctors Without Borders (Médecins sans frontières), with whom she served in Iraq, Afghanistan, and Colombia, in the hope of making the world a better place for everyone.[45]

During one of Elia's missions to Khost, Afghanistan, she came closer to nature despite being in a secure compound to protect her and other foreigners from Taliban attacks. Living in a confined space encouraged Elia and her colleagues to be creative and make use of the garden, where they grew fruits and vegetables, and planted gigantic sunflowers: "Despite the fact that it was an arid region, the plants and everything we sowed were very good, so being there gave us a lot of peace and tranquility."[46] Because chiles are popular in both Afghan and Mexican cuisine, she used the local chiles to prepare Mexican salsas to serve at mealtimes with her colleagues, sharing a piece of home.

When she was returning to the compound after a shift at the hospital, Elia recalled: "There were Mexican marigolds (*Tagetes erecta*) that caught my attention. It was really nice that every time I came back from the hospital, whether it was in the morning or at night, the flowers greeted us because it was like coming home again."[47] The Mexican marigolds were particularly nostalgic for Elia because this is a bright orange flower that is harvested in late October and used during the Day of the Dead celebrations in her home country.

Another inspiring person I met was Paty Fuentes, who had spent more than forty years doing urban gardening. Her home in a residential part of Mexico City was full of lemon and lime trees and a plethora of plants, creating a visual and olfactory sensation. She told me about her experiences in Japan in the 1980s, where she discovered trash separation—a rather forward-thinking concept at the time—and the value of composting, which had helped her to achieve excellent results in her garden.[48]

I really enjoyed the months I spent traveling around Mexico City and hearing people's experiences, as well as learning about new urban gardening techniques such as water-capturing systems using recycled bottles, and generally optimizing space in the spirit of the rooftop farm I had seen in Brooklyn, all of which help to reduce carbon and water footprints. I discovered that even by having a tiny green space available, insects can thrive and attitudes toward the environment can improve. In spite of being one of the most polluted cities in the world, in Mexico City people were able to use their balconies, rooftops, backyards, and other available spaces to grow food and, as a consequence, increase the available habitat for insects and birds.

My mood improved. I felt happier. I even began swapping seeds with my neighbors. One lady I interviewed gave me cotton seeds that, to my surprise, flowered and produced gray-colored cotton eight months later. I recall the joy I felt when one of my neighbors gave me purple beans from Puebla, which produced intense red flowers that attracted three different species of hummingbirds.

It was probably at that point that I realized how important photography would turn out to be. And a few other events further into my journey, particularly the bees that I would discover in Germany and Uganda (chapters 10 and 11), would push me further in the direction of capturing images and spending more time in nature, eventually fusing the two and driving this book forward. My interest in global issues deepened at the same time as my awe for the natural world slowly blossomed and would reveal many of its beauties as my journey unfolded.

Reflecting and Reviving

BERLIN BEES

$\mathcal{A}$fter my travels across the East Coast of the United States and Europe, I moved to Berlin because my former partner already had a base there. I exchanged the warm, chaotic, and intense metropolis of Mexico City for a seemingly more tranquil and verdant city. Berlin also has a rich and diverse population, where German, English, Arabic, and Turkish seem to coalesce and infuse in some kind of linguistic cocktail—another draw for me to move there.

One night, my former partner called my name, frightened. I explained that there could be non-life-threatening reasons for why she had started bleeding, although my heart sank when I foresaw what would come. We rushed to the hospital, where we were directed to the *kreissaal*, or labor ward, where, on arrival, I could hear the screams of a woman freeing life from her womb as if she herself were being killed.

Meanwhile, our examination room was silent as we looked at the inauspicious monochrome colors emanating from the forty-two-inch screen hanging on the wall. The gynecologist desperately tried to find a heartbeat by pointing the ultrasound device in different directions; alas, she couldn't. We were crushed, and I told my ex that I wished I could have taken the physical pain for her. I remember crying so much that night that my tear ducts weren't able to release any more pain. That's when I experienced drought for the first time.

The stresses and strains of the relationship, compounded with our loss and an unresolved issue, concluded in a breakup. I capsized and was lost at sea. Eventually, I decided that I should stay in Berlin and begin to plan the next steps of my journey, which had been put on hold because I had pursued the dream of starting a family with someone I loved—nothing had been more important than providing the right environment in which to receive the baby who never arrived.

Searching for light, literal and metaphorical, to help me emerge from that dark tunnel in my newly adopted city, I was lucky to find an apartment with a south-facing bedroom attached to a balcony. Spring came, so I began to plant sunflowers (*Helianthus annuus*), Chinese lanterns (*Physalis alkekengi*), field marigold (*Calendula arvensis*), and other seeds, all with regular soil without peat,[1] causing an explosion of colors on my balcony. Within six weeks, many of the seeds had fertilized and, at sunrise and sunset, attracted a hive of activity.

Naturally, the flowers weren't just a food source for bees. The occasional bird would come and feed on the sunflowers. The bumblebees would destroy the bluebells because their bodies are large and the petals small and fragile. I saw a similar phenomenon at the botanical gardens in Berlin, where the violet-colored carpenter bees (*Xylocopa violacea*) had fed on a patch of yellow Jerusalem sage (*Phlomis fruticosa*). When I returned a week later, all of the petals were on the ground, leaving the plants naked and exposed.

It was at this time that I stumbled across a fifty-year-old Vivitar manual macro lens at a vintage camera store close to my apartment. I liked the solid metal build of the lens and wondered what I could photograph. I then recalled a time in Mexico when I had taken photos of a beehive and got stung in the ear because I had approached the front of the hive without wearing protective gear and used the wrong lens for the job. This new lens purchase would thus be an opportunity to get even closer and open the door to a world I didn't know existed before.

I headed to the outskirts of Berlin where an acquaintance, who is a beekeeper, regularly does maintenance on eight beehives set in a beautiful orchard. He showed me some of the basic steps of beekeeping, including how the honey is centrifuged and how to locate the queen. The latter was easy because she had a green sticker on her back and was always surrounded by her faithful subordinates, but generally speaking her extremely large abdomen gives her away.

It took a bit of time to get used to composing shots of honeybees because they move quite quickly. I enjoyed studying the bees closely on my computer at home: the small hairs that come out of their large compound eyes on either side of their head; the oceli, which are three simple lenses at the top of the bees' heads between the compound eyes and behind their antennae; and their tongues, which, when fully extended, are nearly as long as their entire body, enabling them to reach a flower's nectar and pollen.

It didn't take me long to find out that Berlin is home to numerous other types of bees, including the aforementioned carpenter bees as well as bumblebees (*Bombus*). A few months after buying the vintage lens, I decided to get a professional automatic macro lens. It has a very quick autofocus motor that allows me to capture better images of bees and not miss a critical moment by clumsily trying to make manual adjustments. I then traveled the length and breadth of Berlin and documented eighteen species of bees of the more than five hundred known to exist in Germany.

What started off as an initial curiosity about this fascinating world quickly morphed into a passion and love for these animals. I discovered that patience is important to get the right shot. I sometimes have to wait for hours for the bees to come, or need to return to a site several times to get the shot I want. The honeybees that I observed in Berlin seemed to be calm in comparison to their East African lowland honeybee (*Apis mellifera scutellata*) counterparts that I would meet later in my journey, in Uganda and Kenya (chapters 11 and 12), but probably not as passive as the carpenter and other bees that I would see in Uganda.

Bees are highly adaptable and intelligent animals. They are excellent builders and cleaners, and their honeycomb is often a near perfect hexagonal structure: a space for future bees. Through careful housekeeping, which includes the use of propolis (a sticky resin bees use in multiple ways)[2] and cleaning, they ensure that the hive is a safe, habitable place for the colony.

The bees' transfer of knowledge to advise other members of the colony about food sources is thought to be done through two dances. The round dance is performed by bees to alert the colony that a food source is close to the hive.[3] The waggle dance is more sophisticated in that it provides other members of the hive two important pieces of data regarding food sources that are farther away: distance and direction.[4] This information essentially gives the other bees a coordinate where the food is.

A European honeybee feeding on a sunflower in Berlin, Germany. Photograph by author, 2019

In addition to taking photographs, over time, I began to take notes of the bees I was observing. While taking photos of European honeybees feeding on the extremely small white thyme flowers at the botanical gardens one afternoon, I saw a yellow European hornet (*Vespa crabro*) fly behind one of the bees, which was less than half its size, and overcome it in less than a few seconds. Initially I felt sorry for the bee, but then I remembered that this is part of a food chain and, therefore, life itself.

Later, I discovered that honeybees can surround hornets and overcome their aggressors by roasting them alive using their collective body heat,[5] adding yet another reason I was in awe of these wonderful creatures. But the climate crisis, which is bringing increased temperature peaks around the world, has put bumblebees at risk of extinction in North America and Europe within the next twenty years.[6] This is why whenever I take my camera and photograph any bee, I feel a sense of joy that they are still alive and gratitude that I have the opportunity to capture that moment.

I spent more time than expected documenting bees due to the fact that their natural habitats are being squeezed. I made numerous visits to Prinzessinnen Garden, which has a range of hives, from the top bar hives that are easy to maintain, to a treetop hive hung on a pulley that, as I would discover later, are bees' preferred homes amid dozens of rows of kale, sage, lettuce, and sunflowers on the ground, providing them with an abundant source of food. Alas, the garden, originally built on a 0.6-hectare wasteland, had to close to make way for development of an apartment block. The hives were subsequently moved to a repurposed graveyard in the adjacent borough of Neukölln, more than twelve times the size of the original site, where death was replaced with life.

At the end of that summer I went to an international beekeeping conference, Learning from the Bees, where experts from around the world gave talks ranging from time series analysis of temperatures at different points within and outside a honeybee hive, to different ways to manage varroa mite infections (a common problem for honeybees in Europe), to the possible effects that next generation mobile (5G) networks may have on bees' electromagnetic fields.

The talk that stood out the most for me was given by Piotr Pilasiewicz, a forest beekeeper from Poland, who uses traditional techniques of sawing lateral crevices high up tree trunks in which bee colonies can swarm and make their homes. He was also campaigning for forest beekeeping in Poland and Belarus to be included as part of the intangible cultural heritage governed by the United Nations Educational, Scientific and Cultural Organization (UNESCO), which was at the final stages of evaluation.[7]

In doing this, Piotr hopes to maintain and preserve a practice that is not only beneficial to humanity's heritage, but also to bees' very existence and a return to their natural habitat, because the honeybees' homes in Europe would likely have been in forests where they would have lived high up in the trees to avoid mammalian predators. Because deforestation has wiped out almost all of Europe's forests, many honeybees live in hives under the watchful eye of beekeepers. The increased urbanization, pollution, temperatures, and the use of agrochemicals mean increased threats to bees everywhere.

THE SHOW MUST GO ON

During this period of my growing interest in and passion for bees I learned about insectageddon, or the mass extinction of insects, that falls within the sixth mass extinction of the species. Insects underpin life on Earth as we know it. The persistent and excessive use of pesticides, fungicides, and other chemicals, together with habitat loss, are key reasons why insect numbers have fallen so dramatically.[8]

During my tour around Berlin, I discovered numerous *insektenhotel* (insect hotels), which are pieces of wood with holes of different sizes that provide various insects, including bees, with places of refuge—particularly useful in an urban environment where nature habitats are getting squeezed. Another challenge is water availability. During the summer heat wave in 2019, I saw large numbers of honeybees gathering at water fountains in Berlin as they fought for survival. That individual bees were able to access water fountains during that excessively hot summer and alert others from their colony not only demonstrates individual skill and coordination to ensure their survival, but is also a sad reminder that the limits of what the environment can handle are being breached.

In Germany, a group of entomologists, through their meticulously standardized methodology, determined that 76 percent of insect species had been eliminated over a period of twenty-seven years.[9] While I was unable to find a comparable study done anywhere else, even thinking back to my childhood in London brought this into sharp relief: I recall seeing butterflies then, but I can't remember seeing one in my adult life there. They had all disappeared, slowly and silently. Only a few people initially spoke out about this, but their numbers are growing, and I am now part of this group of concerned citizens.

In 2019, a study showed that thirty-four different insecticides were found in German pollen.[10] It may well be that other countries have higher levels but simply haven't carried out comparable tests (paucity or absence of data doesn't prove that a phenomenon doesn't exist). It is impossible to imagine a habitable planet if the number of insects decreases further, with the remaining ones absorbing more poisonous insecticides.

Honeybees underpin the pollination of food crops today,[11] through a special relationship with flowers and trees:[12] the reward for the bee is the nectar, and the reward for the flower is the reproduction of its genes somewhere else. Bees are responsible for nearly three quarters of the world's crop pollination.[13] Without them, we wouldn't have as many blueberries, almonds, tomatoes, and so many other foods. An instructive example of the power and scale of honeybees can be found in the United States, where thousands of hives containing tens of thousands of honeybees each are trucked cross-country and delivered to almond and other farms to carry out pollination.[14]

In this sense, the health and future of honeybees serve as a useful proxy to our (*Homo sapiens*) future. If for any reason there is a negative impact on their populations—be it a significant temperature increase causing drought, increased agrochemical use that interferes with the colony's health, or some other exogenous factor—this will, in turn, affect the very existence of human beings.

This realization made me become extremely worried, so I immediately switched my shopping—as far as this was possible—to organic products, which thankfully in Germany is quite easy to do. Germany has numerous organic food shops and, even in bigger supermarket chains such as Rewe and Edeka, organic foods are clearly labeled, making shopping for them relatively easy. They do, however, come at a price: first, many of the foods come from outside the European Union, and second, there is a financial premium to pay.

I sometimes jokingly tell my friends, "Do I want to have money in the bank or do I want to have a planet?" I'm happy to spend 10 euros (US$11) on 500g (around one pound) of organic local honey, which is thick, tasty, and full of goodness, versus a jar of a similar size for a third of the price, knowing that it likely contains a blend of honeys from different parts of the world in a process called transshipping that I would discover later in my journey and may contain GMOs.

The dilemma between buying locally grown GMO foods, where farmers are looked after, versus organic foods that have undetermined working practices, unknown habitat loss, and require lots of fossil fuels to be burned to transport them, such as the organic broccoli from Ecuador I accidentally bought once while doing my weekly grocery shopping, is another quagmire. We have far too many unknowns. As an idealist, I would say that all farms everywhere should make the transition to organic, and that consumers, as far as they are able to, should be encouraged to buy organic and as local as possible, thereby altering the supply-and-demand dynamic into more favorable prices while creating a consumer environment more beneficial for the planet.

———— ∞ ————

The modern era of bees resembles the plight of refugees: forced out of their homes (the forest) to temporary dwellings (hives), limited food and water, and a precarious future, all resulting from the destructive behavior of a largely male-dominated group of people who profit from asymmetries of power to commit atrocities to serve their own causes, such as material wealth, fame, power, dominance, and a long list of meaningless extrinsic qualities that reflect human beings' most unenlightened facets.

It is understood that bees have been present on our planet for at least one hundred million years,[15] whereas humans, *Homo sapiens*, have been on Earth for two hundred thousand years.[16] This means that if the time that bees have been on Earth were compressed into twenty-four hours, we joined them at 23:59:37, making it all the more tragic that we may, in the not-too-distant future, plead guilty to the charge of causing their extinction. It seems intuitive that the bees' success at thriving on Earth for tens of millions of years has to do with their matriarchal society, governance, and a high degree of adaptability, among other things, and ostensibly has nothing to do with *Homo sapiens*. Actually, humankind has done very little, if anything, for bees.

All activities center around the queen bee's well-being. She has the ability to reproduce, and therefore, the capacity to create a new cohort of bees and, with it, the colony's survival. The feminine capacity to create new life isn't limited to birds, bees, and humans. In Arabic, Spanish, Italian, German, Portuguese, and other languages, Earth is a feminine noun. The seed is deposited in the earth and grows there, just as the spermatozoid fuses with the ovum and sticks to the endometrium, and later becomes a fetus. The bond between baby and mother is the strongest bond in a human being's life. In certain cultures, such as the Andean people in South America, life is directly attributed to the goddess of the Earth, *Pachamama*.

Seen through another lens, perhaps why *Homo sapiens* have been so successful during their reign on Earth is the male, testosterone, egocentric fuel. For the man doesn't have a uterus nor a menstrual cycle and cannot comprehend the gravity of life itself, so through his so-called evolution, he cheapens and commodifies what he sees. He steals and kills what he wants, telling his tribe lies to justify his actions. And he wants more even if nature cannot provide, so he resorts to more psychotic violence.

In 2018, the United States reversed the prohibition[17, 18] of the use of neurotoxic insecticides called neonicotinoids, used together with

GMO crops in the country's National Wildlife Refuge System. In the same year, Brazil, which as we have already seen is one of the most important custodians of the Amazon—itself referred to not only as nature's lung, but one of its most important treasures—approved 453 pesticides.[19] Five hundred million bees died in Brazil over a three month period in 2019, attributed to pesticide use.[20] Meanwhile, it is believed that between 1 and 1.5 million bees died in a wine-producing region of South Africa due to Fipronil,[21] a toxic insecticide that is also understood to have caused the deaths of millions of European bees.

Given the level of the existential threat faced by bees, and insects more generally, the role of the beekeeper, be it a traditional, modern, industrial, amateur, or any other type, will be to defend the environment. It won't really matter how much knowledge beekeepers will have amassed if the environment around them has collapsed and taken the insects with it. The bees did very well without us, so why do they need us now? Is the beekeeper's main role today to protect bees from the industries that pollute and eliminate their habitats so that they don't die off, considering that the latter would also mean the end of humankind given their vital importance in pollinating most of the food we eat?

There are an estimated twenty thousand species of bees[22] but no definitive inventory of them. That, in itself, is worrying: human beings are making plans to go to Mars, return to the moon and encourage space tourism, among other far-reaching feats, and yet, we do not have an inventory of every single species of the most important animal in the world. More worrying still, we do not have a concrete plan to protect them, such as the one described in Noah's ark in the Old Testament.

Worse still, imagine all of the behavioral, anatomical, and physiological variations and other special properties of bees that will disappear without us ever having understood them to begin with. This, for me, is one of the darker sides of the climate collapse. Humankind may become extinct in a cloud of its own ignorance. But it is the conscious act of killing and the motive for the killing that differentiates between murder charges and sentences. The universal value to not kill—the cardinal rule of humanity and the one that reigns over all of the Biblical ten commandments, "Thou shalt not kill"[23]—ironically lies within the secret center of the destructive processes that are boiling the planet alive.

The agroindustry and its lobbyists will fight tooth and nail to draw out the mass use of chemicals for as long as possible. Mercifully, groups of people in Europe and beyond do take the lives of bees very seriously. In 2019, in the German state of Bavaria, more than 1.75 million signatures were collected for a "save the bees" petition that, it is hoped, will introduce laws to provide greater protection for bees and other insects through an increase in the amount of agricultural land that meets organic standards.[24]

In addition to buying locally produced organic honey in numerous outlets, the ethical and concerned consumer in Germany can also buy goods in fair trade shops, secondhand shops, and *unverpackt* (unpacked) shops that do not package any of the goods sold, encouraging consumers to bring their own receptacle for their beans, rice, pasta, and so forth. More people, including me, are happy to pay a premium for the privilege of having organic food without wasteful packaging. Similar stores have opened in Amsterdam and other progressive cities around Europe. People can buy a range of all-organic products, from food to toiletries, without having to buy the redundant plastic wrapping.

But all of these greener options are only accessible to people who either have the disposable income to pay the premiums or those who are willing to sacrifice certain pleasures in order to comply with a deep-seated commitment to the environment. For many people in the Global North, and a large part of the Global South, it would be prohibitively expensive to shop at a completely organic, fair trade, and package-free grocery store because they are already struggling to make ends meet. Their financial desperation puts increased pressure on the climate, because those apparently cheap GMO products that come at great expense to the environment are their only real choice.

I regularly have discussions with my friends in Berlin about this, and there is a feeling of responsibility toward and solidarity with the environment and other people. Riding the bike to the supermarket or subway is a perfectly normal thing to see in large cities in Europe. Both behavioral patterns are present, to greater or lesser extents, across the Global North, from the United States to Japan, but it's shocking that only a few countries in Europe, as already highlighted in chapters 7 and 8, are leading the way.

Meanwhile, the rise in hate crimes and normalization of far-right discourse has created the perfect smokescreen to obfuscate the true state of the environment. The climate crisis is inextricably linked with overconsumption, human rights violations, and ecocides. In common parlance, we cannot have a significant reduction in CO_2 when looted oil is so cheap and bombs fall like raindrops, and forests are desecrated while we continue to live life as normal and talk about fixing the climate as if it were in a vacuum. Therefore, human rights cannot be considered a privilege that is conferred on a few members of the human race, nor can such an important value be fully embraced without defending the environment.

A visible and invisible web binds us together. We can see the words, money, and actions that are transacted. What we can't see is the pain or the reality of the other, such as the North Africans who pick fruit in the fiendishly hot south of Spain; the workers in the United States who wear diapers in sweatshops because toilet breaks are prohibited;[25] the bloodshed in foreign lands; and the environment, which is collapsing, because they are all so far away. Their blood and pain are my blood and pain.

In parallel, a silent and pernicious process is the disappearance of insects. We can't see them. They are small, hidden and out of sight, and yet our whole world relies on them. To date, the International Criminal Court (ICC) hasn't held a single trial for crimes against nature or ecocides. Who will defend the bees? Who will ensure that we don't push butterflies to extinction? Will the estimated US$1.45 trillion cost of F-35 fighter jets[26] be able to protect these beautiful animals one day and ensure that forests are not burned to a crisp, that land isn't stolen, and that people can live in peace with nature? Perhaps it might be one good use of those destroying machines.

By the end of that summer in 2019, I had researched enough stories to develop in sub-Saharan Africa, a place I would visit next, which is a critical part of the Global South that is experiencing high levels of population growth. Moreover, given the effects of heat, the region is particularly susceptible to the climate crisis through temperature increases and water scarcity, together with the systematic theft of its resources that has made an already complex and precarious set of circumstances all the more challenging and worrisome.

Although I was still as thin as a rake following my gastric operation, I felt emotionally equipped to plan my journey. I had a renewed sense of drive and energy, and reasoned that I would just need to be more careful than normal. I had no time to lose. I could feel the crackle of the so-called temperature tipping point beneath my feet. I began the process of emailing NGOs and other potential sources for interviews, pored over maps and analyzed different routes, and finally packed my gear.

Ten days before my departure, I called my good friend Daniel Silva, a Portuguese filmmaker I had met in Berlin in 2011 when I began writing my first book. He is another adventurer. After graduating at the top of his class at Lisbon law school, he worked as a corporate lawyer and, like more and more people who question the very nature and purpose of their existence, was happy to trade his desk job for the spirit of adventure and took a round-the-world trip, which culminated in a photojournalism project titled *The World Is Not Flat*.

Today he is a freelance cameraman and editor, as well as an independent filmmaker, based in Copenhagen. Knowing this, I asked Daniel his opinion about a new lens I thought might be useful for my trip. About ten minutes into the call, it was decided: he would film me in Uganda and Kenya for a documentary he wanted to do about my fieldwork and the creative process of writing a book. Finally, two years later than planned, the forces of destiny would carry me to Africa.

Part III

AFRICA

· *11* ·

Flight of the Honeybee

HONEY FARM

Plumes of red dust exploded behind the four-wheel drive's tires as we drove up the steep, bumpy hill, with sixty-foot-high trees creating a slit in the landscape for us to pass through. We were in a rush, transporting a colony of East African lowland honeybees from Kampala, Uganda, to a new home in a newly built farm set inside a forest about an hour's drive north.

We all became concerned when some of the bees started to escape from the hive shortly after leaving the city. I had to open the window to let them out while my host Simon Turner, an affable, laid-back Australian beekeeper who had come to Uganda initially to trade honey more than a decade ago, looked for a spot to stop the car safely on the highway. Henry Tumusiime, his assistant, got out and sealed the newly formed holes with mud that he had in a small plastic bag.

When we arrived at the farm, Simon and Henry quickly put on their bee suits, then carefully moved the hive from the trunk to the apiary. Simon likes to work in pairs. Henry attentively observed Simon making adjustments to the beehive while he smoked the colony with a mixture of cow dung and lemongrass in order to pacify them, a procedure he would repeat during another colony migration and during maintenance work I saw them do over several days.

Even the day before, the colony had seemed quite aggressive and lively, which is why Simon elected to use lots of smoke to pacify the tired, stressed bees, then slowly open the hive to allow cool air to enter.

139

I could see and hear the bees fanning themselves intensely to keep cool, but I could also see that some of the bees on the honeycomb were lifeless.

Suddenly there was a crescendo. More bees had cleared the nectar from their bodies and began to fly away. Henry, being the brave soul that he is, didn't bother putting on the head protection or gloves, enabling him to see more clearly and improve his manual dexterity, and reached into the hive with his bare hands trying to locate the queen. The buzzing sound went from forte to fortissimo when the robber bees of the same species from a nearby colony decided to have a piece of the action and descended on the honeycombs, starting to feed on the free nectar.

I had to take a few steps back because I didn't have a bee suit and was aware that multiple stings is one thing, but being devoured by a fearless group of bees could be fatal. The raiders' bodies were clean and dry, contrasting with the oily looking slick that covered the new colony. Some of the robber bees even licked nectar from their tired brethren. This final observation enabled me to gain a more mature view of honeybees: yes, they are smart and can be virtuous, but they are also trying to survive and, when resources are limited, will take them from weaker contemporaries. In this sense, humans are like bees.

I rushed to the nearby hut to get a bee suit now that the bees were getting even more lively and began circling my head, a sure sign of an impending attack (I'd been told that bees tend to attack faces and necks). By the time I returned, the queen had been found dead. Later, I wondered if this might have happened had there not been global warming: as I witnessed in Berlin, bees are struggling on this hotter planet. During the days that I would spend with Simon, I discovered a passionate, committed human being who cared about his team, the bees, and Uganda.

The bees that I had observed seemed fundamentally the same as the honeybees I had seen in Europe and Mexico: organized, disciplined, and strong, applying their skills to forage for sustenance, build honeycomb, reproduce, look after their young, and die. What astounded me that afternoon was a realization of their potential and kinetic energy, or as I have heard on a few occasions, their "life force." Once again, I felt a closer connection to bees, but this time for the obvious reason that their primary purpose is similar to other life forms, including ours—survival. Then I asked myself, do bees have consciousness?

After lunch, Henry gave me a tour of the apiary, which is divided into a number of sections across the farm. We started off with a real treat for me: stingless bees, which are by far the smallest bee I have ever seen, less than half an inch long. We approached the hive slowly and carefully so as not to disturb them. The stingless bees were very docile, almost friendly; their silvery eyes looked magical and surely contained mysteries that go back thousands if not millions of years. They were especially difficult to photograph because of their small size and extremely dark bodies. I decided not to take too many photos because I was concerned that the shutter noise might upset them.

From there we moved to the main area where six large Kenyan top bar hives were grouped closely together. I learned to walk slowly and not to get close to any of the hive entrances because these bees were nowhere near as calm as the ones I was used to in Germany. A few of the guard bees started bumping into the mesh that covered my face, curious to know the intentions of the new visitor. The sound of tens of thousands of East African lowland honeybees buzzing within their hives, together with the frequent departures and arrivals of foragers, created an exquisite symphony that reflected their majestic life force.

One of Henry's most important tasks is to ensure that the bees are healthy and are not attacked by invaders. Much like a moat that surrounds a castle, each leg of the hive has a protrusion where a small reservoir is filled with just a few drops of oil to prevent invaders such as red ants and termites from climbing into the hive and destroying it.

Henry, twenty-four, who has four brothers and two sisters, mentioned that he was the last-born male in the family: "It has some importance, to be among the last born. . . . When I was in school and out of money, I called my elder brothers or my dad."[1] In spite of the privileges that being both male and a younger member of the family gave him, he assumed responsibility early. Much to the surprise of his neighbors, he was able to look after two hives independently by the age of ten. After studying to be an electrician, he decided to pursue a profession in beekeeping.

"Most of my fellow [beekeepers] see bees as aggressive creatures . . . but when you create good communication with them, you can do everything with them. But there are those which are aggressive."[2] Henry's approach to beekeeping, as was the case whenever he interacted with his colleagues on the farm, and Daniel and me, was to always be softly spoken and gentle: "You try to create a good relationship . . . you have to be calm, do things slowly, slowly . . . So, when you handle those bees in that way, then they will not harm you."[3]

Another one of his rituals is to check each of the hives in the apiary every evening: "I check in on the girls, you know that most of the bees are girls, I say, 'Okay my girls, how are you? I'm going to sleep. I also wish you a good night.'"[4]

A few months earlier, he had found a swarm of bees on the ground when he checked the hives at night. According to Henry, virgin queens don't normally mate inside the hive and, for some reason, the queen of that particular colony was either tired or disorientated on her way back to the hive and didn't enter. He saw lots of drones on the ground who, as expected, died after fertilizing the queen, but other drones were still alive because they hadn't mated. So, Henry acted quickly, helped the tired queen and the surviving drones back into the hive, and was able to save the new colony thanks to his ritual of saying goodnight to the bees. "It's not a burden; it's a passion."[5]

One evening, before dinner, I sat with Simon on the veranda of one of the huts he had recently built on his five-acre farm. We watched the sky turn gray and inhaled the damp air as a deluge of rain descended to the ground, creating small capillaries of water that merged into an artery down the hill, rinsing the surface and making it shiny. The campfire at the bottom of the hill was still alive thanks to a makeshift metal roof over it, held up by some sticks, while the soundtrack of frogs, crickets, and other insects and birds became muted by the sounds of thunder that regularly interrupted our conversation.

Simon's dream is to transform the farm into a research and development center for bees, including a training center where he can continue to run his courses, provide accommodation for ecotourists, and a modest amount of food production. "We're trying to get people to do beekeeping because, instead of chopping down trees, you're planting trees for nectar sources."[6] Because many of his clients are both beekeepers and farmers, they are incentivized to look after their trees, which can be additional nectar sources for bees: "There are different trees which are fast growing, like *Calliandra* and *Grevillea*. They're quite robust. Then you've got other trees like mango. You get fruit on [them] as well, and you can get quite a sweet honey from that."[7]

This vision is too valuable to go ignored, and in this sense I hope that Simon's beekeeping evangelism across Uganda helps to raise awareness of the East African lowland honeybee in the first instance. The major hook here is the financial rewards that bees can provide farmers through greater crop yield and honey sales, and through this encourage farmers to be more aware of their environment, as well as to avoid the allure of industrial chemicals widely known to be detrimental to the natural world.

Simon is trying to create a domestic bee market as a means to fight against the importation of foreign honey that he sees as an existential threat to the local honey business and, consequently, the bees themselves, which could be thwarted by increasing domestic supply. According to Simon, imported honey is low grade and brought in from overcommercialized farms: "People aren't really aware of it if it looks good and it's coming onto the shelves here in Uganda."[8] He also believes that even though the government has imposed high import duties on honey— which is a good line of defense not only for local farmers but also the environment, because goods thus wouldn't need to travel long distances and burn even more fuel in doing so, thereby reducing the carbon footprint substantially—this has certain limitations: "People are dodging it . . . There's always a roundabout way of doing things . . . What I'm seeking to do is to get them to ban the import of honey, full stop, because we can produce a lot of honey here in Uganda."[9]

Indeed, as he went on to say: "We're getting better-quality honey compared with what we were getting in the beginning." He's happy with his suppliers and wants to protect the local market from transshipping: "A lot of it is shipped to Australia, they ultra-filter it, which takes out the pollen." This, according to Simon, removes evidence of the botanical region from which the honey was originally sourced. "Then they'll add some Australian honey so that it gets some Australian botanicals in it, and then it will be called 'product of Australia,' shipped to the Middle East, and repackaged." Simon took another sip of his tea and added, "So, you wonder, 'What's going on there?'"[10]

Even in Europe, some of the labeling is contradictory: you can buy a jar of honey that says it contains both EU and non-EU honey. In other words, the consumer has no idea where the honey comes from, especially for low-grade honey, while the producers (i.e., both the bees and beekeeper), who worked hard at preparing the honey, are anonymized and devalued. This, once again, evokes the dark side of capitalism.

Thankfully, it is possible to buy more up-market locally sourced organic honey, which commands a premium and embraces a triad that I am deeply committed to supporting: bee, beekeeper, and the environment. Long gone are the days of adding a generic honey to my breakfast or tea. I only buy local organic honey, sometimes from a distance of as little as five miles from my apartment, even if that means paying three times the price. I do this because there is only one environment, which is precious and priceless. Moreover, taste and texture take a big jump, and the honey is often far superior to the golden syrup from unknown origins I used to eat.

I joined Henry and the others for a cup of tea. We were all squatting because the ground was wet, gathering warmth from the bonhomie and campfire following an intense day in the trenches. "I'm used to stings," Henry told us. "When I get stung, I say 'Oh, I'm getting some dose.' What I can avoid is an overdose . . . sometimes I don't put a second glove on my hand, except if I get two to three stings, then I have to put it on."[11] Henry believed that he hadn't been sick in several years because of his exposure to bee venom.

As Simon packed the single mattress into the trunk of his station wagon where he would sleep during our last night on the farm, I walked to the hut at the end of the property, eager to get much-needed rest before the next leg of my adventure with him. This would take me to the far north of the country, close to the border with South Sudan. The campfire, the lush green surroundings, Simon and his amazing team, Daniel's company, and, of course, the wonderful bees, including the tiny stingless ones that I had never known existed, had made me feel that the world is far richer than I had previously imagined. I was learning. I was happy. I was excited.

OUTREACH

The palm trees, banana plantations, and cornfields of southern Uganda were replaced with pine trees and beautiful marshes that extended as far as the eye could see, followed by flat plains as we approached the Hutchinson National Park. We drove past a sign warning people of an elephant crossing, then stopped for a snack just where a group of baboons, presumably from the adjacent national park, gathered around a rubbish dump looking for food.

Our journey north was part of a mission that Simon regularly carries out across Uganda to encourage trade in honey, as well as provide training to beekeepers who can buy hives and equipment from him. We arrived in Gulu by early evening. There, the surrounding vegetation was vastly different: I saw lots of fields, especially corn, with the occasional plot of avocado or sunflowers. The fertility of the land seemed obvious because the shades of green were so intense, and every color looked almost oversaturated when compared to similar fields I had seen in North America and Europe.

Upon arrival, we were greeted by our hosts. They gave us a plot where we had to quickly pitch our tents and unload our equipment in the fading light, given that there was a power outage. In the meantime, our hosts prepared a delicious bean, pocho (cornmeal), and rice dish for dinner, which we ate while gathered around a campfire. I had bought a tent and sleeping bag in Kampala but couldn't find an undermat, meaning that the next three nights would be less comfortable than usual.

In order to make good progress the following day, we woke up at 3 a.m. and disassembled our tents in the dark, aided by our mobile phones that doubled as torches, because we still had no power. The hectic schedule had worn me down, and I fell asleep on the backseat for part of the three-hour trip with Simon to Agago, while Daniel rode up front and took spectacular images for his film. Over the next three days, we visited Lira, Otuke, Pader, and Kitgum. Progress was slow as we crawled through muddy, rutted roads in Simon's four-wheel drive station wagon that, like many cars in Uganda, was a secondhand imported Japanese vehicle. Due to recent heavy rains, we had to get out of the vehicle and push it when it got stuck in bogs, with villagers coming to our aid.

Simon carried out several demonstrations of his beekeeping work in English, with occasional summaries in Luo by a local man who works at a partner organization. In each case, we would arrive about an hour beforehand and unpack the kit from the vehicles, which included bee suits, rubber gloves, and boots to protect the beekeepers. Once the audience of about twenty to thirty people arrived, Simon introduced the Malaika honey brand he owns and explained about the bee suit and maintenance of the hives. Local farmers were gathering, looking curiously and inquisitively at the spectacle. During the live demos, he carefully opened the hive using a knife to cut away the propolis that seals the top bars shut. He also showed the farmers how to check on the bees in the top hive, and how to clear out termites and other pests that can be detrimental to the colony and their honey, which he recommended doing once a week.

The wildlife was spectacular, especially in Pader, where I saw white, orange, and yellow butterflies, crickets, and dragonflies that had black and brown wings that looked like they were made of lace. During one of the demos, I grew absorbed in the stunning scenery and asked myself if I would ever see such natural beauty again. Simon was distracted by Daniel's camera, and I used the opportunity to drift off into the adjacent fields for a couple of hours to take photos of an extraordinary bee. It was a bit larger than the East African lowland honeybee, but rather than having an orange thorax and an abdomen with black stripes, it was completely gold. I was transfixed, and my love for bees deepened further.

During several of the sessions, I saw people reviewing Simon's *Beekeeping as a Business* book,[12] which was in its fourth edition and was being translated into Luo, which will help to increase its relevance and accessibility in Uganda. In the book, although certain technical terms are unavoidable, he uses deceptively simple prose together with pictures and diagrams to help novice beekeepers understand how to set up their beehives, together with numerous troubleshooting tips. I was pleasantly surprised when I saw that Simon had included tree planting in his book, together with a wonderful illustration to clearly signal why it's so important, thus making fundamental science accessible particularly to some readers who may not already be acquainted with the key principles outlined in his text.

After one of the last sessions, I caught up with a beekeeper and farmer who looked after twenty log hives that he had set up on his own. He had heard about Simon's training session through a radio announcement and was excited to come because he wanted to move to Kenyan top bar (KTB)[13] hives to increase yields. That could "improve my standard of living, like spending [money] on school fees, medications and other personal things," for his family, which includes nine other siblings. This farmer's approach, like that I had learned about elsewhere in Uganda and Kenya, was to maintain log hives without a smoker or bee suit, a brave and potentially perilous task.

In one village we visited, numerous hives had been set up by NGOs but only a few were actually colonized by bees. Given the fields full of lush vegetation that extended as far as the eye could see, it was unlikely that food shortages caused the empty hives. According to Simon, the cause was poor construction and setup.[14] I also wondered if a lack of training could be the case. If so, that would indicate a kind of hit-and-run operation, whereby the NGO delivers a concept to a community but doesn't follow up locally, albeit congratulating itself for its impact work in the Global South.

Such a view was compounded by the words of a local man who spoke excellent English and joined us at the campfire one evening. He spoke about NGOs as being an acronym for "Nothing Going On," and continued: "Sometimes, not always, they have their own interests, so to propagate an NGO's interests means to deviate from policies and exploit people, exploit the government, or put a project where it's not wanted."

The imposition and implementation of useless projects to serve some NGOs' interests reminded me of the arrogance and aggression characterizing several foreign actors who feel an urgent need to subjugate the Global South to the Global North. They often don't work on equal terms with locals, or even with some of the workers whom they have procured from the Global North. I have seen job postings online in the Global North for work in the Global South ranging from those where applicants are expected to pay to volunteer in the Global South or enter low-paid roles, to senior positions that include excellent benefits packages that one might expect at a for-profit company, which is in stark contrast with the realities faced by the locals they are supposed to be serving.

As we sat outside the hut where Daniel and I would spend the night, I noted that there was no running water or electricity. And yet, bungalows were for sale at a gated community in Kampala, Uganda's largest city, with prices starting at US$130,000, which is completely out of reach for the vast majority of Ugandans. Creating these kinds of unaffordable accommodations, super-fast 4G mobile internet, and cheap mobile phones that bring images and music videos from around the world all help to create the illusion that change and modernity have arrived when, in fact, it might only be upper-middle-class people working in transnational companies, or those who own companies themselves, who will ever be able to afford that kind of comfortable, secure lifestyle. I was left wondering what northern Uganda would look like in ten or twenty years. How many farmers would switch from traditional beekeeping using log hives to the modern techniques? What would the stunning landscape look like in the future? Would there still be such a diverse range of bees and other insects as I had seen?

BACK IN THE CITY

Back in sprawling, energetic Kampala, I met with Thomas Ayshford, a British entomologist and businessman to find out more about something that had sparked my interest since I learned about insectageddon: alternatives to agrochemicals. Thomas opened a small blue plastic container and pulled out a brown powdery substance peppered with gold specks of vermiculite used to provide moisture to keep tiny predatory mites (*Stratiolaelaps scimitus*) alive, which are used for pest control. "It's an exciting time in Uganda right now," he commented, "because of what Simon's doing and because of what I'm doing. It's on a cusp. If people don't move to being more harmonious with the environment, they have no business, and there is no export."[15]

Thomas believes that the Ugandan market is being pushed into using less chemicals, "so I'm able to step in with the biological products . . . what we do is we take living organisms out of Uganda, we multiply them, and we put them into greenhouses and into crops."[16] This was the total opposite of what I had seen in Latin America, where crops were being sprayed with industrial chemicals even in the heart of the Amazon rain forest (chapter 3).

An additional complication of the use of agrochemicals, which in the first instance is related to their harmful essence, such as the neurotoxic side effects of neonicotinoids (chapter 10), is a lack of understanding of how they should be used. Farmers will pick a certain brand because they know it and will spray it on any crop in any situation when, in fact, it's a fungicide: "They'll spray it against ants. It's not helpful to the environment or to them,"[17] said Thomas.

By taking a natural approach, Thomas is able to manage pathogens and pest insects, which, in turn, acts as an enabling factor for his clients who want to export: "There is a ban on chiles from Uganda because of a moth called false codling moth (*Thaumatotibia (Cryptophlebia) leucotreta*). It has become resistant to pretty much any chemical, so the only option to look at is biological."[18] Among his clients are Dutch-owned flower farms: "They are being forced by the customers and their owners to move away from chemicals, so they are being forced to move into biological. And we do see there are really interesting results."[19]

Our conversation moved to how Thomas wants to make other contributions. "I want to play a social role as well. Any business can make money through the big farmers, but I also want to return to the small farmers and sell at cost price or zero in order to help out the lady around the corner with her matoke in the back garden, her chiles or her beans . . . that's where I can give a little back to society."[20] It was uplifting to see that a British-born businessman who had moved to Uganda and fallen in love with the country could have a passion for science and a strong desire to contribute to people's success. This was quite the opposite of the dismal, dilapidated NGO-sponsored beehives I had seen in the north of the country, and the human exploitation, deeply rooted in Africa's colonial history, that I would be reminded of once I reached South Africa (chapter 13).

Thomas didn't see that working in harmony with nature and farmers was necessarily altruistically driven, because he believes that the farmers are getting better results with fewer negative effects on the environment: "It would be very nice to see farmers have an increased yield and not have sick children because [they were] spraying neonicotinoids the day before."[21] Through Thomas's grassroots approach, he has seen increased yields without some of the nasty side effects that the agrochemical industry so desperately tries to brush under the carpet. For example, in March 2020 one of the major industry players, Bayer, which acquired Monsanto, was expected to pay a US\$39.6 million settlement to the class action related to the company's product Roundup, which contains the herbicide glyphosate, and which forty-two thousand original plaintiffs claimed had caused their cancer.[22, 23]

After several intense weeks of filming, interviewing, and traveling, a visit to the Entebbe botanical gardens was what the doctor ordered: no interviews, no questions; just me, my cameras, and nature. Purple dragonflies, orange and black butterflies, the occasional drone fly that mimics a honeybee—quite literally, a wanna bee—and of course, more bees, all within the stunning grounds that lie on the shore of Lake Victoria.

Then I came across two carpenter bees (*Xylocopa caffra*). When I saw them, I was amazed by the intensity of the colors. Their heads and most of their bodies were jet black, with the only differentiator being that one had a white neck and the other a yellow back. Both of these bees, like other carpenter bees I had seen in Europe, were huge. They were three times the size of a honeybee, eclipsing all bumblebees I have seen. In spite of their large size, they were extremely agile and precise when inserting their large tongues into the flowers that swayed in the breezy conditions, and very difficult to photograph due to their erratic, unpredictable flight paths.

It seemed like a fitting end to that leg of the trip. I sat on the edge of Lake Victoria after an almighty thunderstorm and pondered on what I was really experiencing. Each and every time I saw a new bee or observed an existing one, I felt like I had taken a step further into a new dimension. The observational powers through multiple eyes. The circular dance. The waggle dance. The pollination. The sweet nectar. The navigational skills. The poise. The precision. The beauty. Oh, the beauty.

While sitting by the shoreline, my mind jumped to an article I had read a few months before my visit about mussels that had been cooked in their shells on a beach in California due to climate change.[24] The world is undoubtedly getting hotter, but I didn't feel that this would be the bees' biggest problem directly, because one of the key differences that bees have compared to many other species is the ability to work together to regulate the temperature of an enclosed space—in this case, a hive.

The ability to regulate their temperature as a group, then to fly off when their leader dies, shows multiple skills: the ability to adapt and to survive. Unfortunately, as humans, we don't have anywhere to fly to if the planet becomes uninhabitable. There's talk about going to Mars, but this, in addition to being very resource intensive, isn't scalable and doesn't prevent us from wrecking another planet. I kept thinking about the bees and wondered what the world would be like if more people fanned together to keep the planet cool.

Daniel showed commitment to this book by taking a leave of absence from his job in Copenhagen to document a part of my fieldwork, together with an abundance of grit by dealing with what was an intense, uncomfortable, and often unpredictable, yet rewarding, journey. Our friendship grew stronger. His company, together with the process of getting to know Simon, Henry, and the other staff at Malaika, as well as the wonderful hearts and minds we met during our travels, helped me feel that the world could be a better place despite the dark clouds I saw above my head. As the first drops of rain fell from the sky, I was reminded of the importance of water, something that would be driven further home to me in Kenya.

Thank You for the Rain

A LONG-AWAITED GREETING

$\mathcal{I}$t was great to meet Kisilu Musya. I had had a video call with him nearly three years earlier and, finally, the moment had arrived. I gave him a hug straight away, and we had a cup of tea and talked as if we had known each other for years. Actually, I already knew some key parts of his life even before I met him. He is the video diarist and protagonist of *Thank You for the Rain*, a feature documentary directed by Norwegian filmmaker Julia Dahr.[1] Set on his farm close to Mutomo in northeast Kenya, the story depicts his trials and tribulations as a farmer, husband, father of nine children, and, crucially, a climate activist.

The film illustrates the fine line that Kisilu walks every day. Unlike his contemporaries in many countries in the Global North, he doesn't have access to farming subsidies, nor can he rely on government support when disasters strike in the forms of drought and torrential rain, both of which present existential threats to his farm, home, family, and livelihood.[2] I saw Kisilu searching for local solutions and even talking about his experiences at the COP 21 in Paris in 2015, where heads of state, including then-president of the United States Barack Obama, also spoke.

Since Kisilu's speech in Paris, CO_2 levels have risen by 4 percent[3] and, with them, global temperatures, meaning that his situation is even worse. "Climate change has brought about an irregularity in timing. We are expecting rain, let's say, in October. To my surprise, it now rains in another month, sometimes earlier, which means that we are not prepared,"[4] said Kisilu. The unpredictable weather means that he has been caught out in the past because he hadn't bought seeds or prepared the soil: "We have to be flexible, to take action. The rain is coming soon. We have to buy seeds, plan quickly, then plant."[5]

His main crops are corn, black-eyed peas, and mung beans (green gram), as well as some fruit trees including mango and papaya. Some are perennials, such as black-eyed peas. "They grow for two seasons, then we harvest them. Sometimes you find that these perennials don't mature because of the extended drought, and die."[6] The unpredictable harvesting times and crop failures add to Kisilu's worries as they essentially mean that if a certain crop doesn't mature for harvest when expected, then not only does he have less to sell at the market, but he also has less variety for his children who need a balanced, nutritious diet. This, in turn, exerts more financial pressure because he has to buy other beans and grains in order to compensate for the nutritional deficit.

I asked Kisilu if he was tempted to take the easy—and dangerous—route of GMO/pesticides/herbicides. His reply was firm: "I, as a climate activist, decided to shift to ecological farming. We have our indigenous corn seeds, indigenous black-eyed peas. Now we are making the sprays by our hands from our trees. We collect it from our planted trees."[7]

As I transcribed the first day's interviews with Kisilu in my hotel room in Mutomu, Kenya, I recalled my visit to Guatemala (chapter 2). I could sense the desperation of the coffee farmers there, with coffee infection risk, water scarcity, violence, and the additional uncertainty of volatile international coffee prices pushing them to sell their beans almost at any price. They conveyed an overarching feeling of c'est la vie, but Kisilu was different.

In spite of clearly dire situations he has faced on his farm for years, together with the responsibility of being the father of nine children, he has been unrelenting in his search for solutions to mitigate the risks of floods and droughts, with his infectious energy and enthusiasm spreading throughout his village, region, and as far away as Norway and France. His concerns were his family, nature, and humanity, which he stealthily masked with his smile. His fortitude is a stark reminder to never give up.

ECO FARM

I stood at the banks of the dried-out river a few miles from Kisilu's home. Beside it, a well about twelve feet wide and ten yards long was filled with water. Kisilu introduced me to his friend, Ibrahim Kioko Ngeke, who had built the well to tap into some underground water.[8] Ibrahim had put up a wooden fence with thorns in it to prevent animals from falling in or contaminating the water through their urine or excrement, and used a pump to irrigate the nine-acre, ecologically friendly farm, which seeks to avoid the use of industrial chemicals such as pesticides and fertilizers. The eco farm also serves as a farming school where local farmers can learn and share innovative concepts.

The water from the well is connected to an elaborate irrigation system made of small man-made ridges in the soil. Through careful planning of this system, it now takes only a few hours for one person to water the field of spinach, instead of more than a day. Kisilu explained that the water, in addition to improving the land's fertility and the insects', birds', and families' sustenance, was also crucial to increasing their income: "We, the parents, pay for our children's education. Their education is not free. We don't have enough teachers in our primary schools. We employ extra teachers."[9] I was humbled and simultaneously flabbergasted.

In the midst of climate change, Kisilu had the foresight and drive to secure his children's education and future. His daughter, now eighteen, moved to Nairobi to start a college degree in agriculture. "If they are educated, they have power. Knowledge is power. They don't necessarily need to work in an office. They can research. They can think for themselves."[10] Ibrahim's eldest child (of six) is in his last year of high school; "he wants to become a scientist because farming is the backbone of our country,"[11] he added proudly.

I moved the conversation with Ibrahim toward how he sees the future of farming in Kenya. "I have noticed an increase in pests since I started farming. I'm afraid that in years to come, that pests will be very serious in farms if action isn't taken. Also, farming is getting more and more expensive. This worries me."[12] Ibrahim wants to see a transition to solar from gasoline, identifying the latter as a big threat because it is expensive and a major cause of pollution.

We took a break from the intense heat and sat in the shade of a group of papaya trees. I looked across at the splendor that could have been the Garden of Eden. The East African lowland honeybees that hummed over our heads, the butterflies that roamed freely, and the smell of fresh compost made me feel like I was on a very special farm. Kisilu pointed to a tree in the distance that held nearly a dozen birds' nests: "We don't frighten them. They help a lot in eating insects. The birds feel comfortable when they're there. So, we are friends with everything."[13]

We continued the tour of the eco farm. Ibrahim showed me the tree nursery, emphasizing its importance: "In ten years' time, there won't be rain if we don't plant trees."[14] In order to promote biodiversity on the farm, and broaden the range of products that can be sold at the market, Ibrahim and Kisilu have planted lots of mango and papaya trees. These trees also act as a natural sponge for water and provide shade on the ground, reducing the temperature there. As they further explained, "In this farmers' training school, we are planting different trees which act as a hub. When they flower, they scare some of the pests away." The versatility of the trees did not end there: they also mash the leaves to extract the sap, dilute it with water, then fill a spray bottle and apply it as a natural insecticide whenever necessary.

Kisilu and I spoke at length during our first and following meetings in different locations in Kenya about the dangers of industrial farming. He reflected on past actions: "A lot of bad things are happening in our bodies and our environment. We used to buy these seeds from shops to realize that we are doing a lot of destruction, also, pest control. We used pesticides which are toxic to our environment."[15]

Later, Ibrahim showed me a few of the beehives on his farm, one of which is essentially a two-foot hollowed-out log, commonly known as a log hive, set about twelve feet high in a tree. In order to collect honey or do maintenance on the hive, he uses a ladder and goes up without any protective gear, which puts him at great personal risk.[16]

We walked to the other end of the farm and saw another hive thirty feet off the ground, but this one had been chosen by the bees themselves, and he had no feasible way to access it. Bees prefer building their hives high up as a natural defense against mammalian predators such as bears and humans. Back on the field, I could see some of the bees hard at work on the tall corn crops, spending as little as three seconds at each of the yellow tassels that dangled at the top of the plant. This reminded me of the vital importance of bees that I had learned about in Germany and Uganda.

As we walked back to the farming school, I could see that it was a veritable oasis of life. It also serves as a test bed where local farmers can experiment with different techniques, such as the irrigation system they developed, and share their ideas and expertise within the community, enabling them to apply their newfound knowledge on their own farms. Such collaboration not only stimulates ideas, but also builds trust and generates much-needed confidence to confront the array of issues that farmers in this region face.

Kisilu and Ibrahim are friends and colleagues. Their fraternity and solidarity were clear as they finished each other's sentences—the stage of a friendship when non-siblings can truly say they are brothers. As a parting gift, Ibrahim gave me a huge papaya from his garden. Because we had already visited his log hive, it felt natural and right to give him one of the photos of a European honeybee that I had taken a few months before in Berlin. I could sense that Ibrahim, Kisilu, and so many others I had met to that point had a deep love and respect for nature. I could feel this passion permeate my very being.

A GREENHOUSE WITHIN A GREENHOUSE

While we made our way to visit a new dam project, Kisilu suggested that we stop at a large greenhouse complex run by one of his neighbors. A staff member at one the greenhouses, Bob,[17] gave me a tour of the site. The first greenhouse was empty. It had hosted spinach that had been heavily affected by fungi, so the team had cleared that away by fumigation a few days earlier. "We are waiting for their incubation period to expire, then we will prepare the beds and replant again."[18]

The second greenhouse had kale that was also affected, and a third had cabbage, which, to my surprise, also had the same infections. The worker pulled one of the cabbages from the ground and showed me the marks of the fungus and pests, namely, diamondback moths (*Plutella xylostella*), that had wreaked havoc and inhibited the cabbage's ability to grow.

I had been to numerous greenhouses in Brazil, Mexico, and the United States, and had never come across such a large infection nor the need for a farmer to carry out such a drastic procedure. I was curious to find out how they managed to get rid of the fungal infection. "So, you're using fungicides, then?" "Yes," Bob replied, "We use fungicides and pesticides." I then asked him whether the chemicals that he used were bad for the environment. "Yes, to some extent they are, but within a contained environment it's better than just spraying them all over."[19]

Bob wanted to take a different approach, moving forward through genetically modified tomato seeds (hybrids) for the greenhouse. In spite of taking this approach, the plan was to continue using pesticides as insects will still enter the greenhouse: "You cannot run and chase them, so you need to use pesticides now to kill them."[20] I asked him if there were less-aggressive solutions to the problems that the farm was facing. "Currently, we don't have natural solutions to it, but with biotechnology we are developing crops that are resistant to such diseases."[21]

I was left scratching my head by this reply. The monoculture, which had chemicals that are toxic to humans and the environment, grown in a greenhouse built in a semiarid field in the northeast of Kenya, didn't make any sense to me. The greenhouses looked out of place in the Kenyan countryside, and I felt that the faith that he and others were putting into biotechnology had been misplaced. I queried him to see if he understood the effects of the chemicals; he replied: "They really do have a negative effect [on] the environment. When the chemicals escape from this place, they pollute the air. Polluting the air means that we are going to breathe the chemicals around this farm. Some cause respiratory diseases that corrode the respiratory tract."

The worker, who was clearly knowledgeable about science, spoke about these collateral effects matter-of-factly. This somewhat disappointed me, because it was as if the knowledge of the dangers didn't seem to awaken a human instinct to pursue less aggressive means. When we left the final greenhouse, I saw two hauntingly familiar backpack spraying kits similar to those that I had seen previously in the Lacandon Jungle (chapter 1) and the Amazon (chapter 3). Each had a 4.5-gallon capacity, which means that when fully laden, it would weigh about forty pounds. I also saw packets of Ridomil Gold (fungicide) and Actara (insecticide).

It was the first time I had seen any of these things up close. I felt like the relative of a deceased victim who could see the attacker up close, and I felt both angry and sad. My involuntary response was to take my camera out of its case and photograph the spraying equipment and the chemicals. Numerous articles link chemical use and a reduction in natural habitats with a significant decline in numbers and diversity,[22] also known as insectageddon[23] (chapter 10).

Although the use of genetically modified seeds can appear to have the advantage of increased yields—and, with that, bigger profits—this occurs to the detriment of the environment. It is to a large extent understandable that many farmers across the world, particularly those I met in the Global South who are under incredible financial strains, may be pushed over the edge and take this route because they mightn't see another way forward. In this sense, Kisilu and Ibrahim are bulwarks that could stem this dangerous trend.

In fact, the chemicals that I saw being used in the greenhouses and beautiful forests in Mexico and Brazil, which are home to species of animals and insects of immeasurable value, follow a similar pattern to an all-out chemical war against a perceived enemy, be it a fungus, insect, or other living species, instead of a calmer organic farming approach to safeguard the future of everything and everyone. We only have one beautiful planet.

What I saw during my fieldwork in Africa and the Americas is analogous to the way in which the weapons industry is employed to identify an enemy, whether based on fact or otherwise, rather than apply diplomacy, negotiations, and tact to reduce, if not neutralize, threats. But the real interest in both cases is money. The agrochemicals industry and weapons industry share that parallel today, a parallel rooted in history.

During the Nazi era, Zyklon B, which had originally been used as a pesticide,[24] was applied as a nerve agent in Auschwitz-Birkenau, Poland, part of the Holocaust that caused the death of six million[25] Jewish people. Following this grisly period of history, knowledge gained in the fields of physiology, chemistry, chemical engineering, and other disciplines was applied to agriculture, which can be clearly seen today to great effect:[26] kill the unwanted elements, and you can have more yield, which means greater profitability.

Instead of spending time to understand and respect the pests' needs and find natural solutions—like Thomas Ashyford, who uses predatory mites to combat pests (chapter 11), or Kisilu, who makes his own sprays from leaves to thwart predatory insects—the farmers in the greenhouses who face innumerable financial pressures fell into the seemingly easy solution, only to find out that it was anything but that.

Rather than continuing the trade of GMO foods at artificially low prices around the world through the power and reach of global shipping and markets (chapter 1)—foods that are killing consumers through obesity, farmers through suicide,[27] and the planet because it can no longer stand the systematic ecocides at the hands of the agroindustry—it is clear that small-scale, peasant farming with fair prices at local markets is the only ethical and environmental solution to protect farmers and the habitats about which they have generations of knowledge. As I had witnessed both in Africa and Latin America, such farmers were at risk due to climate change caused by an increase in destructive human activity.

Mercifully, as I saw in Salgado's project in the Atlantic forest (chapter 3), nature has the capacity to heal and bring life back. This can only be made possible when people take the initiative and turn their ideas into reality, such as Kisilu and the other brave souls who are confronting the biggest threat faced in human history. While an understanding of science may help, humility and just a small amount of awe for the trees, flowers, bees, birds, and the other musicians in the symphony of life might help us to reevaluate what we are doing in order to reduce the destruction of natural habitats.

A BRAND-NEW DAM

Water shortages were regrettably frequent sights during my travels. I experienced them firsthand in Mexico, Guatemala, Colombia, Brazil, and, until that moment, Kenya. Although the intensity might have varied, the essence was the same: water, the elixir of life, was no longer abundant. Ironically, as I interviewed Kisilu, I learned that a chunk of ice approximately five times bigger than Malta had just broken off Antarctica,[28] which would contribute to rising sea levels.

We stood on the parched light brown soil in the middle of the dam, located a few miles from Kisilu's farm, at the end of the dry season. The dam was built thanks to Kisilu's activism and the film that he stars in, which raised awareness of the issues the region faces. The documentary won numerous nominations and awards at film festivals around the world.[29] I walked up to the top to take a photo of Kisilu, who looked like a small speck inside the large crater that has the capacity to hold one million cubic feet of water, or about twelve Olympic-size swimming pools, reflecting the scale of Kisilu's ambition.

Beyond the extrinsic value of awards and notoriety outside of Kenya, his hard work has made an impact close to home. "After the film was completed, it screened all over the world. One of the NGOs was touched, and we started the [dam] project."[30] No faceless bankers descended on his village to lay out plans for unnecessary hypercomplex dam structures. Instead, a grant by the Climate Justice Resilience Fund (CJRF) was used to implement an elegant solution to the drought problems caused by the climate crisis in the form of a dam, which, according to Wanja Emily, impact producer for the film, will also include a resource center that can be used for workshops and the community, as well as four irrigation model farms connected to the dam.[31]

This reminded me of Fabio's water-harvesting project thousands of miles away in the Jardim Ângela favela (chapter 4), albeit on a much bigger scale, but both solutions are part of a bigger framework to address local issues. São Paolo and Mutomo both had very erratic, intense weather patterns. Harvesting rainwater for later use was a solution that both Fabio and Kisilu thought about even though they were in an urban and agricultural setting, respectively. One tragedy of the climate crisis is that those affected most by an erratic climate are in the

Kisilu Musya stands inside the dam at the end of the dry season. Photograph by author, 2019

Global South, who, coincidentally, face more financial hardship and greater uncertainty.

The year 2019 saw the second hottest average temperature ever globally.[32] Although this caused many European farmers problems, they still generally enjoy greater support than their contemporaries who live south of the Mediterranean Sea. Water scarcity has become a recurrent feature that is clearly linked to the temperature increase anomalies I had regularly reviewed since the start of my journey five years earlier. The temperature trajectory is heading upward, humans are using more water than Mother Nature is able to provide, and the weather systems are changing, bringing more uncertainty, stress, and existential problems into the equation. Kisilu is on the frontline of climate change, and, rather than complain, curse, or capitulate, has found the courage to stride forward, like a brave hero in an epic novel.

About two months after my visit, I saw images of the dam that was half full thanks to some recent rainfall. The sides of the dam were bright green, and I wished I could have been there to see it for myself. But sometimes an image is good enough.

CLIMATE HYPOCRISY

Another torrential downpour greeted me when I returned to Nairobi. Gargantuan clouds of black smoke spewed from the exhausts and irritated my nose and the back of my throat during the forty-five minutes it took to travel the last half mile to the cinema where I had arranged to meet Kisilu to watch his film. A panel discussion was supposed to follow the film, but due to a technical problem, some of the panelists, including Katrin Hagemann (deputy head of delegation, European Union delegation to Kenya), spoke to the audience while the staff at the cinema fixed the technical issue.

The deputy head of delegation shared some nice words about the green deal that Europe would like to implement, including the European Investment Bank's energy investments that will be exclusively renewables. What she failed to inform the audience about was that in Europe, the European Union would soon back thirty-two gas projects,[33] forest fires in Portugal earlier that year were out of control,[34] many parts of Spain were—and still are—at risk of desertification,[35] flooding was becoming more common, and no signs show coal being shut down in the next twenty years.

It also felt ironic that the largely Kenyan audience was listening to this overwhelmingly positive speech given that Kenya and several African nations produce less than one ton of CO_2 per capita, compared with more than eight tons in the United Kingdom.[36] It might have been more fitting for the deputy head of delegation and her entourage to fly to London or another heavily polluting city. Each person that they could convey the importance of a green future to in London would have had a much bigger impact than preaching to the converted in that cinema.

Thanks to the Berlin conference that concluded in 1885, the European powers drew up several of Africa's present-day borders, essentially carving up the continent among themselves and helping to set the rules of engagement, which has encouraged the European looting that continues even today. About three quarters of France's power comes from nuclear sources, with approximately 30 percent of uranium imports coming from Niger.[37] Paris, which had to go into environmental contingency measures to reduce atmospheric pollution earlier during my trip (chapter 8), recently announced the introduction of a solar-powered cycle lane in France through a five million euro investment (about US$5.3 million),[38] duping the electorate and the world audience into thinking that the government is fighting the climate crisis.[39]

Then I recalled the sad irony when Notre Dame cathedral was tragically damaged in April 2019 by a fire. Numerous pledges of financial support were received from far and wide to restore the structure to its former glory. An icon had been disfigured and scarred. A few months later, the world was alerted to record fires in the Amazon rain forest due to increased forest burning by farmers. Innumerable irreplaceable species perished, further accelerating the sixth mass extinction, coupled with an incalculable amount of carbon having been released into the environment; and yet, there were significantly fewer financial offers to help rebuild Earth's greatest temple. The result will be more cheap food on the global markets in months and years to come, more special offers at the supermarket as prices keep tumbling down. Mother Nature was on her knees, screaming in pain, and at our service. Silence.

Such climate hypocrisy is consistent with European colonial history, which has dominated, enslaved, and pillaged the content of Africa.

It continues to benefit from an asymmetrical relationship in which one group profits and another is left with nothing. I cringed during most of Ms. Hagemann's speech and wondered if this talk might be another bullet point in a PowerPoint presentation under "community outreach," or a successful deployment of Eurocentric ideals and values during her weekly meeting with her superiors. Whatever the output would be in the eyes of European technocrats, the input was painful to endure.

But the means with which the greenwashing politicians, business leaders, and the media have worked together is quite remarkable. The fact that so many cars are bought, food and water is wasted, and indifference to the pain and traumas endured by the environment and the people who work incessantly in fields, having to adapt and improvise, demonstrates that the campaign to distract the masses from the greatest-ever threat that the planet has faced is par excellence.

By giving people the illusion that they live in a just, democratic society with values such as freedom of movement and speech, coupled with myriad choices and possibilities that allow the individual unlimited growth potential (one of the biggest myths of all), enabling them to be who they want to be and when they want to be, the custodians of power have conferred the greatest gift of all on those who have money: the ecstasy of unbridled consumerism.

Finally, the technical problem in the cinema was solved. It was a joy to watch the film again while sitting beside Kisilu. The drought, flood, and limited resources were challenges that the film's protagonist tackled head-on with bravery and courage. The use of Kisilu's film screening as a platform to indoctrinate a captive audience didn't faze him. His narrative and the moving images he captured spoke volumes about him. Before we departed, I looked deeply into his eyes and saw hope, fear, strength, and defiance. We both know that we're on a precipice.

· *13* ·

Journey South

THE SUBTLE ART OF PLANNING

The final leg of the trip in Africa required more planning than most legs thus far. I had arrived in Lusaka, Zambia, and pored over maps to figure out the best overland route to Johannesburg, South Africa. The route through Botswana is much longer and, without any reliable information about buses into and out of the country, I looked at the only other option: Zimbabwe.

Having crossed close to thirty countries up to that point, and with all the highs and lows I had lived, I felt deeply unsettled when I read that the south of Zimbabwe was on the verge of starvation. Then there was the thorny issue of my British passport. For understandable reasons, due to my country's shameful colonial past, being British with lots of camera gear might attract the wrong sort of attention and land me in trouble. After several phone calls, I found a bus company that offered direct service to Johannesburg.

Then I recalled an afternoon in Mutomo, Kenya, while filming with my good friend Daniel Silva, we couldn't find anywhere to have lunch or even buy food given that we were filming in the middle of the countryside and the closest town was about ten miles away. Luckily, I found some food in my backpack that we had bought in Nairobi, and split it between us: half a loaf of bread that was still edible, a can of tuna (I ate fish out of desperation for the protein and the salt as I was very dehydrated), a banana, and a liter of apple juice.

Daniel had had to return to Copenhagen a few weeks later, making the final leg an additional challenge because I had gotten used to

165

sharing a room with him and enjoying many fun and unforgettable moments, from seeing spectacular waterfalls to being caught out in torrential downpours when we were fully laden with our camera gear, or eating delicious vegan dishes at an open campfire under the stars. The journey helped to bring us closer as friends.

In order to prepare for the trip the following day, I visited a convenience store near the hotel in Lusaka where the staff seemed resigned to the desperately hot environment in which nothing much seemed to be happening. Due to constant power outages, the diesel generators outside were used to keep the fridges and freezers cool, causing a horrible din and foul smell. The bag of raisins I bought was hot to the touch. Actually, all of the items were hot, but in spite of this, I bought one loaf of bread, two tins of beans, a jar of peanut butter, a packet of cookies, and a few bananas. That would be enough for the upcoming trip.

As I walked through the city that day, Lusaka showed clear signs of colonial history and an imperialist present: the righthand-drive vehicles that drove on the other side of the road, with English being the official language[1] thanks to the British influence. In addition to this, China's mark could be clearly seen in hotels, banks, casinos, and other signage in both English and Mandarin. Perhaps the frequent power outages and intense heat took me back to Nicaragua (chapter 1), causing an insidious feeling to grow inside me with the belief that many people in both places seemed to have almost given up and accepted the hard, daily grind as inevitable as sunrise the next day. Population growth and increased consumer expectations of electrical goods such as fans, air-conditioning, refrigeration, and other tools to keep people cool and food fresh, in countries that are struggling to adapt to change, reflects how strained the global power machine is: there simply isn't enough energy to meet demand in its current state. It's not just Nicaragua and Zambia. It's practically every country in the Global South I visited, whereas the Global North is immune, because the systematic theft of oil, together with the dirty coal business, keep voters happy. But how sustainable is that?

I woke up before dawn the following day and met Karen,[2] the co-owner of the hotel I was staying in, who was lovely. She had taken a couple to the airport at midnight, and now she was driving me. There we were, at 5:30 a.m., having a heart-to-heart while driving through a peaceful, quiet city. I felt sorry for her situation. Her husband, who had gotten drunk in the bar the previous night, was probably still fast asleep. I asked her to help me find an ATM because it was unlikely that I'd be able to buy a ticket at the station with my credit card. In total, we visited seven ATMs: four didn't have power, one didn't have money, and the penultimate one didn't accept my card. In the end, I was lucky and managed to withdraw some cash.

We arrived at the bus terminal that, like others in Africa, Latin America, and elsewhere, can be a very dangerous place in which one is very susceptible. It's obvious to everyone who the tourist is, and that can present traps and other potentially volatile situations of which I prefer to steer clear. Karen helped me battle through the hawkers selling power banks and phone covers, money dealers, beggars, and touts trying to sell tickets that they didn't possess. We found the ticket office and, luckily, the bus to Johannesburg had a couple of spots left.

I gave her a hug and a tip and thanked her. A few minutes after I took my window seat, money changers offering very poor rates came on board, opportunists who assume that people don't understand when a rate puts them at a big financial disadvantage. There's nothing like a captive market. A few hours later, once the bus departed, a pastor stood up and delivered an impassioned oration in which he invited us to pray together, after which he collected money from the faithful passengers.

"Your visa receipt, please?" asked the Zambian immigration officer at the Chirundu border post.

"Ah, I've got the stamp; it's in the passport," I replied.

"I need the receipt for the visa!"

"I'm sorry, but I wasn't given a receipt. I didn't know that I needed one."

Then a snowball of questions and answers meant that I had to get out of the line and take a taxi a mile back to the main Zambian border office where officials deal with complicated cases such as mine. The bus driver's assistant, a man in his early twenties, helped me find a taxi, but it was still being used by a money dealer making a foreign exchange transaction of US\$13,500 with a woman in her forties in the backseat.

Once their commercial business was done, we drove back across the Zambezi river and headed to the main immigration office. What followed was a situation that is extremely unlikely to happen very often in the Global North for someone carrying a valid passport and visa, because it's normally possible to cross borders with relative ease (of course, the same isn't true for refugees, as discussed in chapter 9). I was greeted politely by an officer wearing a clean white uniform (Mr. Nice), an indifferent and apathetic boss who was in plain clothes and spent his time picking his nose while I explained that I was never given a receipt when I entered the country (Mr. Grumpy), and a young woman with beautiful braids and brightly colored acrylic nails who was busy on social media and barely acknowledged my entrance (Miss Social Media).

Mr. Nice was very diplomatic: "I'm very sorry about what has happened. We will need to investigate what has happened to see if other people entered this particular port of entry and didn't receive a receipt."

"I completely understand," I said, trying to maintain a reasonable attitude so that the situation could be resolved quickly, "but I'm here with the bus driver's assistant. The bus is at the border across the bridge, and the other passengers are waiting for me. I have a stamp in my passport."

"Yes, sir, but, you see, this is serious. We need to investigate what happened."

Mr. Grumpy took the passport from Mr. Nice's hand and walked out of the room with his subordinate following him. The driver's assistant started panting and hissing, called the bus driver and spoke in another language I didn't understand, then announced: "We're going to bring your bag here. You can take the bus tomorrow."

This is not the kind of suggestion I expected, especially given that we had only been there for fifteen minutes. I tried to bargain for extra time. He kissed his teeth. I turned to Miss Social Media: "Do you think that you can speak to your colleagues to find out when this will be resolved, please? The bus is waiting, and I need to get to South Africa."

When Miss Social Media said, "Stay here tonight and just take the bus tomorrow," I felt a knot in my stomach. Although I could reason that she was doing her job and following orders, I still felt frustrated and impotent, as well as angry, because I knew that I had done nothing wrong. I imagined that this was the kind of mediocrity that many Africans have to face when there is an asymmetry of power. Then I remembered the Syrian single mom, Saba, whom I had met in Slovakia (chapter 9), and the millions of people like her who aren't able to move freely. Perhaps this was just a miniscule reminder of how unjust life is for others.

Then the driver's assistant rolled his eyes and kissed his teeth for a final time and announced that he was leaving, asking me for my seat number so that he could identify my hold luggage. He promised that he'd send it to me using the same taxi driver who had brought us to the immigration office. I implored him, "Please don't leave me here. I can't wait in this place for twenty-four hours."

After doing breathing exercises to calm down, I looked at the clock on the pallid wall. Another two hours had passed. The bus company hadn't even bothered to return my luggage, or perhaps the taxi driver had stolen it, but that was the least of my concerns. Border crossings in the Global South can be volatile, lawless places, especially after dark. I've crossed many land borders in Latin America and always avoided them after dark because I know that it's much easier to be ambushed when assailants can run off into the night in a place where there is, generally speaking, less security. When I crossed the Kenyan/Ugandan border at 1 a.m., one of the money changers shortchanged me, but a repeat event in Zambia would be trivial compared to my fears of being robbed, assaulted, or worse.

Ultimately, it wouldn't matter how many countries I'd been to or how low key I looked. I had a twenty-pound rucksack full of camera gear that was essential to my work, and I didn't let it out of my sight. I didn't have mobile data, which meant no internet. I had left a bag full of food on the bus and only had a small packet of cookies and a bottle of water in my camera bag. I wasn't panicking just yet, but I could feel the tension return, so I did some slow breathing and took another sip of water.

Time is elastic. It can expand and contract. Time stood still as I sat in that dismal room for hours, waiting for a solution. Then a Congolese woman with disheveled purple hair entered the officer's room. She seemed to be complaining about taxes that she didn't want to pay, but Miss Social Media told her that they would need to be paid to get the entry stamp. I put their dialogue on mute.

Mr. Nice was back, but with no news nor any sign of my passport. I started getting agitated. He kindly asked by me to be patient. Meanwhile, a South African gentleman was handed in by a Zimbabwean immigration officer because he didn't have a Zambian exit stamp in his passport.

"Do you have the money for the exit visa? Look, there is no stamp!"

"I can call my cousin; he can send some money."

"Do *you* have any money to pay for the exit visa?" Mr. Nice's patience had eroded to my levels of despair.

"I can call . . . just let me call someone."

"I will have to arrest you." Mr. Nice calmly took handcuffs out of his drawer, handcuffed the man, and then walked out, leaving me in the room with the suspect, who asked me if I could lend him US$20.

Later that afternoon, I received news: they had determined that I had, in fact, paid for my visa (which was already in my passport), and I was now free to leave. But—there's always a "but"—how would I make it to Johannesburg? I had previously decided not to push my luck in Zimbabwe, but it didn't make sense to backtrack either. I closed my eyes and thought that I should make enquiries about getting a bus back to Lusaka, because I knew that a room would be available at the hotel I had stayed in earlier and Lusaka was far less unsafe than Harare. So, I decided to return to Lusaka and at least get some rest and food, neither of which looked likely to happen anytime soon.

Then I saw a bus with a sign for the Copperbelt (the northern region of Zambia), which meant that at the very least I could get back to the outskirts of Lusaka, then take a taxi to the hotel. As I took my seat, I asked my neighbor how long they'd been at the border crossing: "We've been here for seven hours. One of the passengers has a problem with her papers."

Some men in the back were getting impatient and started shouting. In the absent person's defense, one passenger turned around and announced, "We came together. We will leave together!" His spirit of solidarity resonated with me and a few others. About thirty minutes later, the Congolese woman with purple hair boarded the bus. She started shouting in a mix of French and another language I didn't understand to the men at the back of the bus.

The result of that day's voyage was that I returned to the room I had stayed in in Lusaka about twelve hours after I had vacated it in the morning. The ripped curtains, allowing what remained of the day's light to illuminate the room were yellow and faded, showing tenuous signs of a glorious past. When I had checked in on the first day, Greg,[3] a white gentleman in his sixties and co-owner of the hotel, explained that load shedding (scheduled power outage) was in place in Zambia and that the shower could be operated using the diesel engine that supplied electricity to the rooms during specified times. Greg was born in Northern Rhodesia in the 1950s during colonial rule. He made it a point while proudly showing me around the rest of the hotel that he is Zambian by birth but, in fact, British.

A calendar outside the room listed the days and times of planned power outages (i.e., the load shedding). The Kariba Dam, which straddles Zambia and Zimbabwe, is the engine room that powers both countries, but the unseasonably low water level in Lake Kariba meant that power needed to be prioritized to the Copperbelt region in the north of the country. One of the biggest ironies of climate change is that even though the ice is melting on the poles, which is increasing sea levels and drowning islands, I consistently saw water shortages in Mexico, Guatemala, Colombia, Brazil, Argentina, Kenya and, up to that point, Zambia.

While drinking fruit juice at the bar in the evening, I watched the local men and women having drinks with some of the other foreign visitors. Greg entered the bar with a glass of whiskey in one hand. He grabbed one of the local men by the face, placing his thumb on his guest's left cheek and three fingers on the right, smiled at him and asked, "How are you, my friend?"

Later, I asked Karen a question that had been nagging at me for a while: Was Greg, her husband, ever violent? I tried a more subtle question: "Is your husband mean to you sometimes?" Her voice modulated: "Sometimes he tells me, 'You don't fucking do anything! You're lazy!' But he doesn't see the work that I'm doing. Last night I finished at 1 a.m. It's 11 p.m. now. I will sleep for only a few hours because I have to take you to the bus station again tomorrow."

Karen is too kind and generous for her own good. I had to tread carefully. "I'm speaking to you as a friend, not as a client: you need to look after yourself . . . you're a young and incredible woman!" As I lay on the bed, I reflected on Karen's life. A Zambian-born, black woman, she was working herself to the bone. Cooking, helping guests (like me), and generally ensuring a tight operation while her husband seemed oblivious to her hard work, care, and devotion to their business. I asked myself how many Karens there were in Zambia, Africa, and the world. Of course, the sad answer is that too many women are subjected to all sorts of unfair treatment at the hands of men.

If men, who historically have been the captains of governments, armies, industries, and commerce, cannot treat women as equals, then how could we possibly imagine these same individuals will treat the environment with the care and respect that it deserves? Life itself is communion. So, if the laws of nature, as evidenced by the threats faced by bees, trees, and other forms of life, are transgressed, then the only logical conclusion is a reaction that has already begun: the sixth mass extinction of the species. My hope was and will always be that we can mitigate some of the harm done through love, education, dedication, and perseverance.

BREAK FOR THE BORDER

On my second attempt to leave Zambia, the bus station looked very different. This time, I arrived there before dawn: entire families slept outside the ticket offices, using cardboard boxes as mattresses and blankets that presumably they'd brought with them. The sun peeked over the top of the buildings, slowly turning the colors in the sky purple, orange, then finally sky blue. Everyone was awake. The ticket offices' shutters began to open, vendors started selling coffee and tea, and finally, the manager of the bus company arrived. He honored my original ticket and called the driver's assistant of the bus that I had traveled on the day earlier, who confirmed that my luggage was on its way to Johannesburg.

Upon departure, Peter,[4] the driver's assistant this time around, announced, "Before any introduction, let us close our eyes and pray." I didn't close my eyes, but I listened carefully and respectfully as Peter asked God to lead us safely to our final destination. After a resounding "Amen" that brought many voices on the bus together, Peter went on with the administrative side of things: toilet breaks, border formalities, and breakfast, which was fried chicken and French fries.

I finally settled in and took advantage of the vacant seat next to me to stretch out my legs and listen to some music. As the bus approached the border, the roads became narrow and twisty through some hills. I looked out of the window and saw the semiarid landscape where occasional shoots of green came out of the branches of some of the lucky trees. I closed my eyes and expressed a wish for rain to come soon. Then

I recalled the previous day and feared both the exit from Zambia and the entrance to Zimbabwe.

Miss Social Media was at the exit desk and started laughing when she recognized me. Not wanting to be mean spirited, I went along with her joke. I had slept for three hours, it was a boiling hot day, and my normal amiability seemed to be evaporating but somehow clinging to me like the sweat on my T-shirt. "I hope that you don't say anything bad about Zambia." She entered my details into the system and, a few clicks later, stamped my passport. Finally.

I walked over to the Zimbabwean border post. No computers in sight. All of the forms were duplicated using carbon copies. It was like being in a living museum. I went to the adjacent counter to pay for my visa. The officer in charge asked me how my time in Zambia had been. "It was hot, but OK," to which he responded "Yes, been hot here too. No rain!," then lamented that he didn't have five US$5 dollars' change for the $60 I had handed him. He began to smile when I offered no resistance to losing the $5, which he was almost certainly bound to have in one of the busiest border crossings in southern Africa. He looked at me and chuckled: "I'll get myself a drink with that."

At another counter, a more junior official who presided over stamp duties carefully input the information from my application form into a visa that would be stuck into my passport. Unfortunately, when I boarded the bus, I realized that the border official had entered the second and third digits of my passport number the wrong way around. I had a small feeling that I might have an issue when I reached the border with South Africa.

I closely followed the route on my mobile phone, which had GPS. We avoided the capital Harare and often took unpaved and unlit roads that felt desolate. Whenever the bus passed a town or city, I saw long queues at the gas stations. Even Rwanda, Kenya, and Uganda had had bus stops where it was possible to use a restroom, however unhygienic, and buy food and drinks; but, for some reason, the bus driver on this trip didn't stop at any bus stations, which I assumed must have existed in Zimbabwe. Before sunset, the bus stopped briefly in the middle of the road to give passengers (male and female) an opportunity to take a restroom break in an adjacent field.

My intuition told me that several risks might accompany stopping at a bus station or in one of the towns in Zimbabwe, from police bribes to armed robbery or some other form of violence (something I was well aware of while traveling in Latin America), to locals who might be desperate to flee the country and might attempt to the board the bus. Whatever the reason for not stopping in any of the towns, what I had read in news reports made me feel deeply concerned about the state and fate of the country: water had been recently cut off from Harare,[5] doctors had been on strike for several weeks due to poor pay (US$200 per month) and equipment,[6] soldiers were receiving a third of their rations,[7] and two hundred elephants had died due to a drought.[8]

At night, every town we drove through was pitch black, with the exception of homes or businesses that had generators or possibly solar power/batteries. What I saw was unlike any other place I had been on the trip so far. All of the towns looked run down and unloved, requiring a lot of effort to imagine the country's former glory when it was called Rhodesia, a time when Queen Elizabeth II was the head of state,[9] as well as head of the commonwealth to which the country belonged.

Suddenly, a wall of orange streetlights appeared in the distance, growing as we approached the South African border. On arrival, queues were very long and disorderly. My body was tense as I handed my passport over the counter. I tried to regulate my breathing and appear calm as the border official passed it to his colleague, who flicked through it with indifference. It was close to midnight and another influx of people entered the immigration hall. Finally, I got my stamp.

I breathed a sigh of relief and walked in the dark toward the South African border where the queues were a little less chaotic than on the Zimbabwean side. Once in the main building, I saw some of the border agents playing video games while others were browsing the internet on their desktop computers and thought that they might have been on their break. Passport now stamped, I walked to the bus where two South African immigration and customs officials checked each passenger's passport and hand luggage.

One Zambian passenger complained that he got slapped on the face by a border official because he had only been in Zambia for a day, which was obviously a border run. Another Zambian got arrested for having some kind of issue with his passport. Given its mineral wealth and relatively good infrastructure compared with its neighbors, South Africa attracts economic migrants in a similar way to Mexico, which absorbs Central American migrants looking for a more secure and prosperous life, or to Western Europe, which appeals to Eastern Europeans. Regardless of geography, people want better opportunities and greater prosperity.

The whole border fiasco had cost me a day of my trip and extra taxi rides and hotel costs, not to mention the added stress of only having a few hours' rest in Johannesburg after the twenty-five-hour bus ride from Lusaka because I already had interviews confirmed for midday. But my discomfort was trivial compared to the gargantuan issues I observed in Zambia and Zimbabwe. While there were greater semblances of modernity in Lusaka in the form of infrastructure such as banks, hotels, and highways, both countries face an almighty problem with the water shortages that show no signs of abating. This is combined with the increased population growth, which will mean that Zambia's estimated population of 18 million people in 2019 could grow to anywhere from 36 to 42 million by 2050, according to UN population forecasts.[10] It's unfathomable to imagine how doubling the population would change the face of the country and how it might be possible to cover people's basic needs given the extraordinary pressures that are being exerted on the environment, as evidenced by the lack of water.

· *14* ·

Generations of Miners

DEEP UNDERGROUND

"*My* day starts at 3:30 a.m. Shower. Breakfast. Arrive at the mine at 4:30 a.m., change to underground gear. Meet my supervisors, arrange the job for the day, then I go underground,"[1] said Victor (not his real name),[2] a miner with more than twenty years of experience working in different types of mines across South Africa, including gold and chromium. I had secured a meeting with him on Sunday, his only day off. We stood on the sidelines of a local football match in Rustenburg, a mining town about eighty miles from Johannesburg.

Victor had numerous responsibilities at work. He managed two stoping[3] and two development teams, who drill new ore and do maintenance on equipment, respectively, bringing his headcount to about forty staff. The hot and dusty environment underground is highly tense because an incorrect calculation could mean death for some or all of his teams. I was surprised when Victor told me that his job felt like slavery. I asked him whether the term "slave" was too strong given the history of involuntary movement, incarceration, sale, and death of people, clearly different from the circumstances he and the other miners whom I had already interviewed faced.

"It's not a strong word to use according to the job that I am doing because you are there for more than 10 hours a day. That's slavery to me."[4] Although his contract states that his working day is eight hours, he often works for ten or more hours from Monday to Saturday. I challenged Victor, suggesting that he was free and did have choices; his reply was firm: "I have to comply. I have six children. They depend on me. They need to go to school, shoes, clothing, everything. My eldest is twenty-six. I still support him."[5] He also supports his wife, who looks after their younger children at home, meaning that his 30,000 South African rand monthly salary (about US$2,000) barely makes ends meet.

The mining environment that Victor and the other miners I interviewed described paints a loud, hot, and precarious underground world in which they knew that the risks were high but, due to a lack of opportunity, felt compelled to enter. Some had shoulder problems following years of overuse. Choice, it seems, is relative.

In Bekkersdal, I met with Richard,[6] a thirty-year-old former gold miner. He had held numerous jobs before his mining career and then, at the age of twenty-five, found a job as a gold miner. "There was a time when I was going to work, I was working [the] night shift. He [my son] was crying. . . . He didn't want me to go to work, and I was like, 'No, I'm going to work, I have to go to work.'" When Richard arrived at the mine that morning, he entered the cage lift even though it hadn't been serviced and was squashed in with nearly sixty other men in a space certified for forty. Suddenly, the cage went into freefall down the shaft; he recalled this terrifying moment: "My stomach dipped . . . we were stuck there for two and a half hours." In that moment, Richard considered himself as good as dead. "I told myself, *this is over, this is the end of me.*" He started to think about his family, "what I know is that when you die in the mines, there is money that they give your parents or next of kin." He believed that any financial compensation would never replace him, and that any accompanying letter that would be sent to his wife would have been insincere because he felt cheated and used by the mining company.

His average monthly salary was 7,500 South African rand (about US$500), which, when discounting his food and transport costs, his wife's fees for training to become a teacher, and his son's school fees and transport costs as well as food, meant that he was left with nothing. After three years of hard, committed work, he was made redundant.

When I met Richard, his wife was pregnant, meaning that his responsibilities would increase even more, but he didn't seem fazed by this. He expressed his belief in the young people who live in his community and the desire to steer those who use drugs onto a different path. "I used to play the tuba. We used to have a band . . . I want to start an orchestra where they can play clarinet, saxophone, pianos, violins."

These words instantly reminded me of the good work that so many other musicians are doing in South Africa and elsewhere in the world, such as Paraguay, where lives are being transformed through music (chapter 4). Even though we may have a less than ideal set of circumstances, how we decide to maneuver them is up to us, even if our options are limited and require a lot of hard work. What made me feel honored to have met these and the other miners I interviewed was their love for their families, the commitment they show in risking it all each time they enter a mine shaft, knowing that they may never resurface, and the bravery to speak their truth.

⎯⎯⎯ ⚭ ⎯⎯⎯

William[7] greeted me warmly at his home, a modest ten-foot by six-foot improvised dwelling in one of the less developed parts of Soweto. He offered me something to drink, then sat on the edge of his bed and began to tell me about his mining career, which started in 1985 and ended in 1990. He had worked laying the railroads for trains to transport gold out of an underground mine. Back then, he was married with children and, like most of the miners I had met, told me that his wage wasn't enough to make ends meet.[8]

As a supervisor, those who reported directly to him complained about their low wages and asked if he could speak to management to negotiate a pay rise. "I told them, 'If you speak about money, you are going to be fired.'"[9] Despite this apprehension, William and the team spoke to the bosses and were promptly fired. They were then immediately replaced by another team, ensuring no disruption to the track operations, which were fundamental to the mine's success.

William also wanted to show me his pay slips and the original work permit card he had during apartheid. I was simultaneously appalled and curious when he showed me the dark blue card, nearly fifty years old, listing the companies he had worked for. I felt a bit sad when I saw his original black-and-white photo, in which he looked like a strong, handsome man. And there we were, in his home many years later, after he had been unemployed for nearly half of his life.

We placed all of the pay slips on his bed and sorted them by date before I photographed them. Then he showed me the termination letter; this appeared to be inconsistent with the pay slips, which William asked me to clarify. I had never seen a South African pay slip before and said that I didn't know why there was a difference. He asked me to look into this for him. I accepted because it was the right thing to do, but I did caution him that I couldn't guarantee that I would be successful.

As William spoke, I looked at the ceiling—it had cardboard installed as insulation—and the assorted strips of wood that comprised the back wall of his house. "I'm using a gas stove here," he explained, "If it will burn here, I am also going to burn, that's for sure."[10] How could honest, hardworking people be treated like this? What had changed in the South African mining industry since the fall of apartheid if the laborers are still black men working in a dangerous environment with nonexistent job security? Everyone whom I interviewed was on a contract, meaning that severance could be instant. This is probably the reason why the United Kingdom and other countries have adopted zero-hour contracts. It keeps people on their toes, and a long line of desperate people is waiting outside if they don't like it. This is what capitalism has become in the twenty-first century.

The interviews with the miners were difficult. Many of those men had battered bodies and crushed souls. Most of them agreed to the interviews on the basis of anonymity. It was as though working in the mines was their only way of supporting themselves and their families living in townships (cities deliberately built for the black South African population during the apartheid era). Many of the men I spoke to didn't have a formal education, but some did. Some spoke several languages and had studied at university level, but the lack of upward mobility reduced their possibilities to working in the mines, a job that in South Africa seems to be dominated by black men from townships.

Although most of the men I interviewed were angry, tired, and disillusioned, there was one man, Moeketsi Baloe, who had worked in the mines for nearly fifteen years whose experience and outlook were vastly different. After graduating from high school in Soweto, he completed a two-year motor mechanic apprenticeship, then decided to try his luck by traveling to Rustenburg, "the city of mines." He landed his first job in a platinum mine through a friend.[11]

In spite of the job being physically and mentally draining because of the heavy equipment and the inherent risks and intensity that come with working inside a hot, dangerous mine, Moeketsi enjoyed it: "We joke around just to make it comfortable for everybody."[12] The ambition and foresight that took him out of Soweto to search for opportunities further afield pushed him even more: he retrained to do engineering maintenance inside the mines. Today, he is the person who ensures that all of the machinery in the mine works. Without him, the mining operation would stop. Even though he has a more senior role, which includes line management, he likes to get stuck in and to help others (i.e., he isn't a generic leader who doesn't have domain knowledge). His passion and joy are infectious, his calm a unifying force.

Moeketsi dislikes the idea of doing the same thing over and over for a long time. Age thirty-five at the time of interview, he had already gained experience in different roles, but he didn't want to stop there and grow old in the mining town. He dreamed of opening a resort in the future: "I intend on studying architectural technology next year. My dream is to do different things. Hopefully, one day, when they combine, they will mean something."[13] Moeketsi clearly demonstrated his strategic thinking and his willingness to invest additional time, money, and effort to make his dreams come true. I felt encouraged by Moeketsi's outlook, which reminded me of my friends living in the Jardim Ângela slum in Brazil (chapter 4), as well as Ibrahim and Kisilu at the eco farm in Kenya (chapter 12), among other resilient and courageous people I have interviewed and observed over the past nine years around the world.

Moeketsi's outlook also reminded me of the adage that education sets people free: it helps individuals reimagine their lives, recognize others, and make them more employable, among other factors, regardless of whether they live in poverty, are refugees or displaced people, or simply want to improve their lives. Of course, race, religion, gender, class, nationality, and some good fortune play a part, but education is arguably the ultimate passport that enables us to envision new possibilities and find the means to pursue them.

Most of the miners I interviewed, regardless of their role, age, experience, or the particular kind of ore they were working on, were unemployed. This made me ask myself what South Africa will look like once all of its metals are exhausted. Many of the interviewees referred to former mining towns as "ghost towns." With an increasing population, the next question is how these communities will adapt to life once the mines have closed, and I suspect that Moeketsi and Richard hold an important part of the answer: education. Even as a university student in London, I recall a professor who showed us the results of research he had done in Australia using a bacteria that could metabolize the walls of disused gold mines, creating a sludge that could be siphoned off to extract practically every last atom of gold in it.

But such solutions don't respond to the deeper problem. Mining is driven by an insatiable appetite that is part of a temporary and disposable culture. The phone that is replaced once a year. The extra television or tablet computer. The new car in the driveway (even if it's electric). We wouldn't need to put these people's lives at risk and harm the environment if we consumed less, recycled, and upcycled more.

An additional concern I had was when I learned about the high degree of variance in the South African miners' monthly income, which went from as low as 7,500 South African rand (US$500) up to 40,000 South African rand (US$2,700). It didn't seem that the financial compensation, including for those on the higher end of the pay scale, was sufficient or fair given the risk, time, and mental and physical stamina required to do such intense work. Also, these men were mining gold, platinum, chromium, and other metals that form part of an extremely lucrative trade for all sorts of industries, from jewelry to electronics, meaning that there is no legitimate reason why they couldn't have been paid well or looked after properly. How much does a Rolex or Cartier watch cost? What is the price of human dignity?

EXTRACTION AND INEQUALITY

I spent most of my time in South Africa with my friend Nimrod Moloto, whom I had interviewed six years earlier for my first book. His inside knowledge of Soweto (South West Township), Gauteng (which means "where there is gold"), and other provinces, were indispensable in gaining access to many of the interviewees, as well as a better understanding of the context. We went on long walks together and discussed the state of South Africa and the world, and caught up on each other's personal lives. Nimrod told me that tons of gold leaves South Africa and is unaccounted for, helping to ensure the asymmetry that was pain-fully clear to see.

Even though apartheid had ended, it seemed that the 9 percent of the population who are white own most of the land,[14] and every single miner I met wasn't white. The barbed wire and electric fences, keeping gated communities at bay from attackers in South Africa, also didn't seem a world away from many affluent neighborhoods in Latin American cities. They did, however, seem completely different from the pristine, calm Park Avenue and Hampstead areas in New York and London, respectively.

On the final day of my visit to Soweto, Nimrod offered to take me to the Avalon cemetery where Lilian Ngoyi, a political activist who rose up the ranks of the African National Congress (ANC),[15] rested beside Helen Joseph, a South African originally from the United Kingdom who dedicated most of her adult life to fighting the same injustices.[16] Together, they led a march of twenty thousand women to protest against pass laws that required all black and mixed-race women to carry identity cards.[17]

I had spent several days in Soweto, which made me feel very happy given the warmth of the people. Nevertheless, the recent interview with William who lived nearby, and standing with my friend Nimrod in front of the tombstones of the two great activists who had truly tran-scended, added a mix of melancholy and introspection to these positive emotions. I tried to imagine what life in Soweto had looked like during apartheid, notably with regard to the violence, oppression, and other injustices that these two women confronted, which ultimately drew them closer together and made their union stronger.

Since their tombstones were covered in dust and dirt, Nimrod suggested cleaning them, using some pieces of newspaper we found lying on the ground. Afterward, we stood in front of the graves, closed our eyes, and observed a moment of silence. My heart felt warm, and I felt closer to Nimrod. A tinge of happiness entered me because I was reminded of some of the shared values that characterize a good friend-ship such as empathy, integrity, and solidarity.

Although the inequalities I saw in Africa resembled those I had seen in Latin America, they seemed to be more intense and prolific. In both regions, a lack of infrastructure in many of the places I visited—including basic or no sanitation, temporary structures such as tents or pieces of plywood, cardboard, and metal cobbled together to make permanent habitable structures in the same city as luxury homes, malls, and international hotel chains—made me feel disgusted and angry.

The notion that fair trade will help lift Africans out of poverty is a well-intentioned but misguided idea at best. Buying fair trade organic coffee from a farm in the DRC, Uganda, or Kenya, or installing beehives but not providing adequate training or maintenance, as was the case with a group of NGOs in northern Uganda (chapter 11), and other similar practices create the illusion that assistance and support are being offered. Although such endeavors may help some of the locals, the scale of the problem is far too entrenched and systemic to be fixed with superficial gestures.

About the time I was interviewing the miners in South Africa, I read that a submarine was carrying out deep-sea mining tests in the Mediterranean Sea. This, if successful, would help increase the supply of cobalt,[18] which is used in electric car batteries (chapter 6). In 2014, the UN's International Seabed Authority (ISA) had issued licenses for exploration work.[19] Dredging the seabed will disturb an already fragile ocean ecosystem whose CO_2 levels have increased, making it more acidic, with less available oxygen and higher temperatures, among other worrying aspects. The mining in question will almost certainly be to the further detriment of the environment and, ultimately, everything and everyone.

Meanwhile, child labor is known to be used in the DRC to extract lucrative cobalt ore,[20] which the submarine in Europe would so desperately wish to extract in offshore mines to help the clean and green emission-free vehicles that will grace the pretty, pristine avenues of the Global North. Another source of child labor in the DRC is the extraction of coltan, which contains niobium and tantalum, both prerequisites for cell phones. Coltan is exported through a long, convoluted process that sometimes involves shipments through numerous countries[21] before it arrives inside a cell phone somewhere else on the planet. The carbon footprint: incalculable. The human impact: unimaginable.

At sixty miles long, the world's largest conveyor belt is located in Morocco and helps move phosphorus, which is essential in agriculture, out of Africa,[22] African food is also exported, even though there is famine or near famine within the continent, which has been the case for decades.[23] Such a perverse situation is a result of greed, corruption, and mismanagement. The acceptance of this as being part of an unfortunate aspect of life reveals how apathetic views have become. We cannot expect equality when systematic theft of resources from practically every country on the continent exists. The rate of mineral extraction means that for the most part, those raw materials will run out within a few decades and, with them, important revenue sources.

Africa, like Latin America, in addition to being rich in minerals and culture, is endowed with extremely diverse ecosystems that are at risk. The mass extinction of species underway will mean that the processes that create the exquisite tapestry of variety of color, beauty, splendor, and natural defenses will continue to disappear.

Meanwhile, and as already discussed at length, the climate is warming and becoming increasingly unpredictable. Desertification, the logical outcome, I have seen progressing with my own eyes. The effects of climate change mean increasing uncertainty with respect to rain patterns and temperatures, which, in turn, means extreme weather in the form of drought and floods. Seasons are becoming less clear cut. A combination of all of these forces may possibly encourage farmers to cross the line and use more GMOs, insecticides, and herbicides, thereby creating a vicious downward spiral to destruction through submission.

The attitude of the Global North toward Africa is far more nuanced than that toward Latin America. In the latter, many Europeans mixed with indigenous people, creating *mestizos*, literally meaning "mixed," which is why, being mixed race myself, I am able to blend into the background in Latin America. In Africa, however, the colonizers, especially the Belgians, British, French, Italians, Portuguese, and Dutch, kept the power, money, and their gene pools to themselves, and only on rare occasions would buck that trend. The Global North's relationship with Africans is not dissimilar to its relationship with the environment. It is out of touch, exploitative, takes but doesn't give, is based on preconceived ideas of entitlement, and drenched in outright greed in which winners and losers are clear.

In this globalized and increasingly connected world, illusions of infinite growth and choice have been perpetuated by social media:

lifestyles that can furnish us with an ever-increasing number of options that confer freedom, entertainment, and joy unto us is the biggest myth in *Homo sapiens*' history. Meanwhile, mediocrity and popular culture have perpetuated a singular definition of beauty as being Caucasian. From Mexico City to Johannesburg, even the most economically disadvantaged individuals have cell phones that give them access to this elusive and illusory world; and many, now that they have seen through two-inch screens what the Global North looks like (opulent, white, amazing), feel that they should have something similar.

But it's the needs and wants of a privileged few who own the agriculture, defense (attack), banking, technology, media, and other industries that have the capacity to bring humanity together or to fragment it. The global banking system and big tech, where nouveau riche billionaires decide what we should see on social media, and faceless investors who trade people's futures, can get rich quickly without much effort.

The violence in some parts of both Latin America and Africa is terrifying, yet their distance from Europe means out of sight, out of mind. The genocide in Darfur, Sudan, which began in 2002, has been taken to the International Criminal Court.[24] The Central African Republic also experienced ethnic cleansing in 2014, but those who were killed were Muslim and Africans, so it hardly received any media attention. Boko Haram kidnapped 276 girls in 2014.[25] Al Qaeda are active in Algeria, Mali, Mauritania, and spreading farther in every direction. Silence.

The DRC is a particular case in point. The country has had numerous infectious disease outbreaks, resulting in the deaths of five thousand children from measles (a known and preventable disease) in November 2019, while the world's second biggest Ebola virus disease (EVD) pandemic in 2018 claimed more than twenty-two hundred lives in the DRC,[26] helping to reaffirm the country as a mysterious, dangerous place. Ebola, as I would learn later, is related to deforestation (chapter 16). Further images of doctors in white overalls, face masks, and gloves demonstrated how dangerous, contaminated, and unknown the country and, by consequence, continent are. Meanwhile, many people sitting at home in the Global North silently accepted that they had no way to help those people. It's such a pity. And yet, there are no issues buying more mobile phones and electric cars whose batteries so heavily rely on cobalt sourced from the country.

What is even more remarkable than the deaths that received so much media attention is that the humanitarian crisis caused by violence in the DRC has resulted in more than 4.7 million displaced people,[27] an unfathomable number, living in unsafe and precarious conditions. This was another underreported story. Why? Because in the Global North, an African life is not considered to be equal to a European, American, or Japanese life. And, besides, if the reporting actually indicated that the violence is linked to mining, then perhaps people might think twice about replacing their mobile phones every twelve months.

For the brave few who make it to Africa, lots of fun can be had: bungee jumping and selfies next to Victoria Falls, a safari to see zebras and elephants up close and personal, white-water rafting, thousands of dollars a night to camp under the stars and be served by an African butler to evoke colonial times, hunting using real rifles and bullets (a pursuit the former king of Spain enjoyed),[28] or haggling at a bazaar in Morocco. Then come the pictures of visitors who pose with small African children who look stunned and bemused in their holiday snaps. "Oh, and the people are so humble. So simple. So nice." Of course.

But why are Africa and its people so misunderstood? Any signs of African leadership that has the temerity to stand up to the Global North are quickly dealt with. Muammar Gadhafi was eliminated, but I am still at a quandary as to his crime, given his cordial relationship with Italy and the philanthropic work he did in Africa. Today, Libya has a large presence of terrorists; meanwhile, Europe has secured yet another source of oil there, which removes the expense and transit time of oil from Saudi Arabia and the Gulf, and with it, provides some of the belligerents fighting in Libya with leverage when procuring oil from other export countries. In any case, what happened in Libya is a reminder to other African countries not to mess with Europe, the United States, or their allies.

Compliance and obedience are staple ingredients needed to maintain power in Africa. The rules are simple: you can pillage as long as I get a small cut. With the exception of Libya, this century hadn't seen a major intervention in Africa. This may be seen as a tacit agreement between the former colonies and their masters, but the African people have paid the price.

A hungry, disorientated, and dysfunctional continent with limited possibilities is exactly what the Global North wants. Some find a way to leave, either by boat across the Mediterranean Sea, or exceptionally talented engineers, doctors, and scientists who contribute to health services, industry, and academia in lands far away. Others simply cross borders to try their luck in a neighboring country, all of which add to the entropy and instability of the continent.

POPULATION FALLACY

Much of the discourse around climate change, recently upgraded to climate emergency, centers around a mystical number of years we have before positive feedback mechanisms—that is, a snowball effect—will kick in. They will cause further temperature increases that will, in turn, induce further ice melt, sea level rise, and potentially release methane stored deep in the oceans, among other risks, all of which will further accelerate global warming and take us to a point of no return. Too many concurrent processes at work, including agrochemical use, insect decline, water scarcity, deforestation, and population growth, to name but a few, so it is impossible to set a precise date for this tipping point.

An important factor to consider is that Africa is witnessing a population explosion: by the year 2050, the medium variant forecast is that sub-Saharan Africa will have nearly 2.1 billion people,[29] about double the 2020 figure. Such growth is clear to see in Uganda's main city, Kampala, as well as other cities such as Nairobi, Lusaka, Kigali, as well as smaller towns I visited during this leg of the journey: it's a young region. Many people have smartphones, and electricity is starting to reach even some remote villages, thus putting more expectations and pressure on families to comply with modern modes of living.

Such a scenario would put this region, already struggling to cope with just over a billion people, under an increasing amount of strain and, therefore, opens up a myriad of questions. How will the current and future populations be fed, housed, and provided with water, sanitation, electricity, education, health, social welfare, and national security? How will the agriculture industry and infrastructure keep up with future demand when it is clear that countries such as Zimbabwe are on the verge of famine and total collapse? There doesn't appear to be a plan. It's a life of excess and joy for those who live off other people's hard work, resources, and misfortunes, but for the vast majority of people that I could see, life is quite the opposite.

Economists would say that, overall, poverty has been reduced in Africa. Epidemiologists might say that infant mortality has been reduced in some regions, and sociologists could argue that the middle class is larger. But these types of relativist views have a fundamental flaw: Africa as a continent is poor in absolute terms even though it has trillions of dollars' worth of natural resources, a disparity largely due to systematic theft of its resources even while the planet is being cooked. The saddest part of all is that the people of this continent will suffer the most even though they have done nothing wrong.

To better understand how Africa arrived at this point, one has to go to the 1884–1885 Berlin conference at which Africa's borders were carved up (as alluded to in chapter 12), creating agreements deciding who could steal from where.[30] It is a precedent that carries to the present day. From lavish palaces and stately homes, to museums filled with loot, to ivory keys on pianos in people's homes, Europe is still awash with stolen diamonds, gold, artwork, and other African artifacts. It is, therefore, ironic that one has to pay money to see some of these exhibits, and more ironic still that one can be arrested if one steals stolen loot and returns it to its country of origin. The concept of ownership in this context is bizarre and confusing.

China, which did not participate in the Berlin conference, isn't interested in putting troops on the ground (yet) like its contemporaries; instead, it uses its big checkbook and invests in mega projects, using state-owned banks, construction, and other companies. Throughout my trip, from Argentina to numerous countries in Africa, China's presence and influence became clearer. The manufacturing superpower sources copper from Zambia (and even bought a mountain in Peru for its copper),[31] coltan ore from the DRC, and countless other metals to help it consolidate its world dominance in manufacturing.

The notion that foreign individuals, governments, NGOs, or any other group of people should come to Africa and exploit, indoctrinate, coerce, and pollute, reveals several strange and perverse attitudes and behaviors, as alluded to by the majority of miners I interviewed in South Africa and the gentleman I met in northern Uganda who didn't trust NGOs, among other things. First, it assumes a kind of intellectual osmosis at the beginning of the educational process, where knowledge is passed from a greater concentration (mainly American and European) to a lower concentration (African) through a porous membrane called indoctrination and humiliation. In the Global North, such practices, as I will explain, are acceptable on the moral grounds of helping the ignorant and less fortunate, or a greater cause that is the precious and important lives of people in the Global North.

Second, this notion assumes that there are too many Africans and that women's rights will help to reduce this number. Speaking on the subject of the great risk of overpopulation in Africa and supporting the idea of imposed birth control under the guise of women's empowerment, French president Macron said, "I always say: 'Present me the woman who decided, being perfectly educated, to have seven, eight or nine children.'"[32] Actually, one of his colleagues, Ursula von der Leyen, the president of the European Commission, has seven children,[33] and countless other high-profile and otherwise successful women who by choice—and by right—have had three or more births. However, Europe has no plans to limit births there in the systematic way European governments and high-net-worth individuals envisage for Africa. They look at African lives as if they were a bacterial or fungal growth on a petri dish that needs to be disinfected.

Clarifying the motivations to impose birth control on Africa a step further, the Danish minister of development, Ulla Tornaes, announced that her government would earmark 91 million Danish kroner (US$15 million) for family planning, "to limit the migration pressure on Europe, a part of the solution is to reduce the very high population growth in many African countries."[34] This normalizes the view that more Africans in Africa are an existential risk, not only to Denmark but also to Europe, as they might need to escape, so a preemptive strike is justified: birth control. Further, the use of the word "solution" in Tornaes's speech very eerily echoes the Nazi term "final solution" (Endlösung), which ultimately led to the extermination of millions of Jews (as discussed in chapter 12). If, indeed, she sees a continent with too many Africans, will Ms. Tornaes's wish be to execute a more definitive, or final, solution to the problem and kill Africans en masse?

It is quite easy, therefore, to draw a parallel between Denmark's military activities in Iraq, Libya, Syria, and elsewhere with this clear government statement. It indicates that sovereign states can do as they please, that no one can stop their bullets or the "offer" of hormones that disrupt women's menstrual cycles—a natural process—in order to stop life. Once again, the hypocrisy of European values allegedly based on the Bible is clear: the most important value is to not kill (which is why homicide is illegal in all European countries), but it is only applicable within Europe's borders, especially if white people are the victims. Europe is united in its violence and consistency. Rather than cause this body of countries shame, however, it acts as if it is perfectly normal and acceptable to behave this way.

Third, consider the perpetuation of the myth that foreign education can emancipate Africa's people. Indeed, as I have seen on my travels in Brazil, Uganda, Kenya, as well as during research for my first book,[35] it was patently obvious that education undoubtedly enables people to better understand their world and to ameliorate their often difficult and constrained living conditions. However, in most cases, individuals find their own solutions without having foreigners impose their values on them, the form of education alluded to earlier.

In cases where NGOs were involved with the young musicians I interviewed for my previous book, they acted as facilitators. That is a lot different than the brainwashing variant of education espoused in Africa. To be clear, I have the utmost respect for workers on the ground, such as Tatiana Rojas, who risked her life and sacrificed her family time to support a vulnerable group at the coal mine in Colombia (chapter 2), or a French woman I met who had malaria twice when she worked in the DRC and didn't abandon the village because she was so committed to the humanitarian mission. I know personally or have read about countless similar people. What arguably merits less respect and more querying is the way in which some NGOs come to Africa (and the Global South in general) and carry out their projects.

Moreover, the educational paradigm is often anthropocentric. It values the human as the central gravitational force in the universe whereby all things are at its service, such as water, food, air, and raw materials. This is one of the biggest falsehoods in education. Human beings are just one element in an indeterminable and extremely broad range of brushstrokes on a beautiful canvas called life, which is quickly fading but, mercifully, can be remediated or possibly even restored to something closely resembling its former glory if the right circumstances prevail.

Finally, the Global North's brand of education is enshrined upon an asymmetrical, selective distribution of facts to perpetuate fairy tales of entitlement. While children in the Global North go to school in automobiles made of foreign raw materials, burning foreign fuel, using computers made from foreign raw materials, their packed lunch full of foreign food, clothes made in sweatshops, and an unstoppable supply of goods creating the illusion of abundance, their parents' argument seems to be that "we're the ones who have the brains and the know-how." Instead of going on an educational crusade in Africa, why don't the NGOs and their sponsors, governments and individuals alike, focus their efforts on educating the mining, automobile, agricultural, defense, and other industries to really understand what's going on in Africa, to protect Africans from the theft of the continent's resources and collateral effects of all the chemicals being sprayed across it?

Perhaps the same NGOs that have ideas of grandeur to save Africans from themselves might do well to travel to the United States, the United Kingdom, Spain, and other countries and refresh people's knowledge of the African slave trade, the abolition of slavery, the end of segregation, and innumerable acts of violence and intimidation, to get to the point where the Black Lives Matter movement still hasn't made a dent in a clearly rigged system that makes black Americans five times more likely to be incarcerated than white Americans.[36] People are murdered for no more reason than the color of their skin, such as George Floyd, whose final words before he was killed by Derek Chauvin, a white police officer, were, "I can't breathe."[37] He couldn't breathe, and Africa can't breathe.

The Global North, which, as we have seen, wields enormous power through capital markets (chapter 5) and violence, can exert a similar choking force to further distort an asymmetrical balance of power between it and the Global South, which it needs for food and natural resources to maintain the standard of living that the North has been accustomed to for centuries.

Such a distortion of power and wealth has created the conditions to derail Africa further by brainwashing talented farmers, who have passed down their old practices for generations, to become subordinate and indebted to industrial farming based on GMOs and industrial chemicals, and to apply their numeracy and literacy skills to understand how much debt they have and how they can pay it off because their situation is so desperate: acquiesce or die. A parallel to this in the Global North may be the many university graduates who rack up tens of thousands of dollars of debt and have to take any available job because their situation is so desperate after graduating. In both cases, the narrative of submission is key to perpetuating the neoliberal policies of growth and greed, which go hand in hand.

But Africa's problems cannot be blamed entirely on foreign meddlers and oppressors. The bureaucratic mediocrity I witnessed in Africa and detailed in my tribulations in Zambia (chapter 13), and to a slightly lesser degree in Latin America, are additional inhibitors to change. That, coupled with violence (state sponsored or otherwise), lack of social scaffolding that impedes mobility, together with theft, also paints a rather bleak picture for the youth of sub-Saharan Africa at a time when this region is undergoing a population explosion.

Killing and stealing go against universal values that transcend religions and times. And yet, these are widely accepted each time we buy a computer, cell phone, or fill up our automobiles with gasoline; bloodstains are on all of them. Nowhere in the Global North's crusader discourse is any mention of the simplest way in which poverty in Africa could be addressed: by instigating an end to the systematic theft and pillage of its resources, together with a fair price being paid for their purchase.

The climate crisis is most likely to affect this beautiful continent and its great people more than the rest of the world. Witnessing the paradox of Africans who sit on an incredible abundance of wealth and yet are, for the most part, economically disadvantaged, made me reflect on another strange situation that I faced: feeling joy and motivation at documenting the incredible work of Kisilu, Ibrahim, Simon, Henry, and the other people I met during my journey, but also feeling terrified. I knew the precipice was there. I could hear it creaking with every footstep. But then, I recalled that I live in the Global North and that my purpose would be to share what I had seen so others could make more informed choices. Every single person can be part of this change.

Part IV

ASIA

· *15* ·

In the Shadow of the Towers

ILLUSIONS

*S*uddenly, loud Middle Eastern music began playing from the huge speakers dotted around the complex. That caused everyone to gravitate toward the huge fountain with dozens of nozzles that began spraying water tens of feet up in the air, creating the largest dancing fountain in the world in front of Burj Khalifa, the largest tower in the world,[1] adjacent to Dubai Mall, one of the largest malls in the world. Then, the tune to *Mission Impossible* began playing. More lights and water acrobacy. A visual and aural tour de force was in full swing.

Lots of parents held up their children to get a better view. Some grabbed their selfie sticks. Many others tiptoed and stretched their hands into the air to record the festival of light, sound, and water on their phones. I saw heads and torsos move from left to right, following the choreographed show, and lots of smiles. It was how I imagined Disneyland might be. Being in Dubai, I wondered whether I might see fire-breathing snakes setting a million thousand-dollar bills alight as the air began to cool as the evening sun disappeared behind the skyscrapers ahead.

Curious to see exactly how much choice those grand malls offered, I went on a mission to see how many sources of honey I could find. I discovered varieties from Greece, Turkey, Germany, Yemen, Pakistan, and Oman in one supermarket (another world record had probably been broken). I'd never seen such a wide selection of honey in one place. As I walked the aisles, I could see people cramming more and more food into their carts, water from Fiji and France, cookies from the United Kingdom, and tea from India, as if it were an Olympic event: competitors from Latin America, Europe, Africa, the Middle East, and beyond, trying to consume as much as possible. The food was as cheap as in mainland Europe despite the fact that it had crossed seas and oceans thanks to international shipping, which is cheap and dirty (chapter 1).

In addition to a remarkably diverse array of food from around the world, the supply of goods was almost infinite, ranging from electronics to luxury brands such as Cartier, Chopard, and Chanel, whose shop signs were in English and Arabic. An ice-skating rink in another mall. Constant advertising for a new luxury apartment complex or investment opportunity. Together, the kaleidoscope of images, textures, and colors helped to create the illusion of crossing the bridge between East and West, traditional and modern, conservative and exotic.

Moral judgments aside, many things I had seen up to that point lacked any semblance of style or class. The spectacle was mind-blowing, breathtaking, and disgusting all at once. New York, São Paulo, Buenos Aires, and Mexico City seemed to be pedestrian compared with what I would see during my time in Dubai. I was nearing the end of my trip. I had already visited twenty-nine countries on this journey and had seen the huge efforts being made to reduce carbon, develop solutions, and reimagine the world. But here I was simply astounded by the consumption and lack of consciousness. Surely most of the visitors must have been aware of the climate crisis at some level?

After having time to recover from the sensory overload, I headed to Al Shabkha, a neighborhood in the north of Dubai, to see more of the traditional trade of Persian rugs by Iranians. Other Middle Eastern, Pakistani, and Indian traders sold low- to middle-end jewelry in what might look authentic to the disingenuous visitor in Dubai's *souk* (gold market). The first salesman I met showed me some "Italian Silver." He punched numbers into his oversized 1980s calculator and tried to sell me some tacky jewelry. Naturally, I couldn't possibly buy anything that more than likely had come at the cost of the environment and human dignity (chapter 14).

I entered a half dozen shops and saw more Italian, Thai, Turkish, and even Indian designs. Yet, no one could give me a clear-cut answer as to which mine the precious metal came from. In all likelihood, they just didn't know, due to an intricate web of theft of metals from both

Latin America and Africa that I had begun to unearth following my visit to South Africa.

Many, if not most, of the men I saw trading likely had left their homelands to come here to make a modest living through trading. I saw them rushing around, buying and selling precious metals from goodness knows where, speaking in languages I could not understand, with the occasional tourist stopping to take a photo of one of the shops or talking to one of the many salespeople trying to entice people in. I was completely lost.

A WORLD AWAY

The early morning sun shone on the old, decaying buildings in Satwa, casting long shadows across their facades, while in the distance hotels, marinas, yachts, trading floors, and malls bathed in modernity. A small neighborhood around a mile away from the city center, Satwa was different in many respects from the artificially cosmopolitan city center. It was as if I were in a different city.

When I took out my camera to take photos near a café, a group of men of South Asian origin quickly left their coffees and headed in different directions. An African man who was passing by waved and said, "Hey, nice camera." "Hey, man," I smiled back, and gave him a fist bump. He continued walking. I clearly looked out of place there. Few outsiders go there, in the same way as it's uncommon to see tourists in the projects in New York or in Tower Hamlets in London. Although those places exist and have immense value because people call them home and they are part of the city, they are invisible in the visitors' guidebooks, maps, minds, and hearts.

As I walked through the neighborhood, I could sense a community atmosphere. The pace of life was slower. People were riding bikes. I saw numerous bike repair/sale shops, kind men working at the bakery, many apartments with clotheslines laden with clothes hanging out to dry, cheap haircuts, fifty cents for a cup of tea or coffee, a few dollars for lunch. Some men were having lunch and drinking tea; others sat in the shade, holding hands. A group of Filipino men and women, gathered around the adjacent table outside a shop, were smiling and laughing during their conversation. Almost everyone I saw looked happier than those in the malls, who were trying to look perfect in the increasingly ubiquitous and plain global fashion scene. If I were ever to return, I'd probably visit Satwa, a slice of Lahore and Manila.

Because some people spoke neither English nor Arabic, I wondered how the different communities would be able to interact. Bill flyers for shared rooms stuck on street corners clearly indicated the desire for certain tenants based on race and or religion. Indian only, Muslim only, African only, Kabayan (Filipino countryman) only, bunk bed, single room, twin room, and so on, together with the price and phone number. The best one I saw was: "any kind person."

Apart from the African man who had greeted me earlier that day, it seemed to be a real challenge to connect. Given my linguistic limitations of not speaking the locals' languages, together with no outside help, I had to resort to the most important of human forms of communication: smiling and sign language. I felt warmth from those with whom I could not speak. I looked into their eyes to try to imagine where they came from and why they were so far from home.

As I walked on, a regrettably familiar sight lurked in the alleyways—numerous prostitute cards littered the ground. This kind of commodification of women instantly reminded me of Amsterdam (chapter 8). I was saddened further when I learned that even if a sex worker is a victim of human trafficking and has been involved in this trade against her own free will, then rather than be treated as a victim of two atrocious crimes,[2] she would face punishment because prostitution is illegal in the United Arab Emirates (UAE). Such a strange form of justice, if it can even be called that, helped me realize that humanity and compassion are not values that can be bought and sold in a market. Those values come from a deeper place through which an empathic evaluation of the other's suffering affects us to the point that we cannot accept injustices and we embrace and support those who most need it.

SKI SLOPE

I tried to have my regular video call with my friend Daniel Silva, who had filmed me in Uganda and Kenya. We normally use a popular video call application, but the call dropped after a few seconds. We tried a few other applications, all in vain, and then I went to the Emirates Mall, near Burj Al Arab, a luxurious hotel built on an artificial island, to see if perhaps the internet there might be better than where I was staying. I had the same problems. It was unclear why the calls had been blocked, especially given that most of the population are foreigners, many of whom are likely to rely on this as a means to stay in contact with their loved ones back home.

While walking around the mall, I stopped for a coffee and saw people skiing in the indoor ski slope. Keeping such a large space at around 0°C in one of the hottest countries in the world is an indisputable feat of engineering and shows the power of human ingenuity. Sadly, it simultaneously reflects another vulgar excess at a time of chronic shortages of water both in the region and the world. If only the engineering could have been applied for some greater good.

The extremity of having a ski slope in such conditions indicated to me that money has metastasized the country into something that doesn't have a higher purpose other than accruing wealth for rulers and investors who want to gamble their money there. It was as if money were a universal truth that ruled above all else. It was difficult to discern the real Dubai beyond the gold market, a trading port, and a few traditional boats.

In a further effort to differentiate itself, Dubai's metro system, the equivalent of the New York subway or London underground, has a "gold" class. But I'm not sure who uses it besides the odd tourist. Silver (economy) class is where Indians, Pakistanis, Iranians, Filipinos, and other foreign nationals, the occasional tourist, and I were packed tightly during rush hour. Even public mass transit had classism. Elitism, like anything in Dubai, is commodified, marketed, and sold.

Behind the smokescreen of the marinas, fountains, exotic sports cars, and exclusive resorts lie many unsavory sides. The UAE has had fighter jets in Libya and Syria for reasons that aren't entirely clear, given that it shares no border with, nor faces any existential threats from, either country. But war is considered sport among some of the world's elite.

A different approach is needed, not just in the UAE, but in other nations that produce and consume large volumes of oil. Namely, that innovation and the only possible future lie in safe, clean, and renewable energy—that is, not the biofuels that are destroying ecosystems in Brazil (chapter 3) or nuclear power (chapter 17). Oil is a black, tarry substance that can easily hide the blood that it has soaked up. After it is distilled and burned, it becomes invisible, so it's no wonder no one has any problems burning hydrocarbons like there's no tomorrow.

Much like the bees in Uganda, which had free, unprotected nectar up for grabs, people have rushed to the kingdom to get their piece of the action. From the executives on six- and seven-figure salaries with full expat packages, to the workers in Satwa, it seemed that everyone was financially better off by being in the UAE than back home, as long as the money kept on flowing. But while interest payments, share dividends, bond coupons, and other revenue streams line people's pockets, the environment has a bigger overdraft.

I recalled the farmers in the south of Mexico who went on strike because of water shortages, the power outages in Zambia because the Kariba Dam didn't have any rain and, of course, Kisilu, who lives in the eye of the storm of the climate crisis in Kenya, where the extremes of drought and torrential rain are becoming more frequent. I wondered how he might have responded had he seen the dancing fountain show, or what the skiers would have thought had they seen such extreme water and power shortages in Africa.

Human rights and environmental protection go hand in hand. We cannot have one without the other. Equilibrium, harmony, respect, knowledge, and compassion: this is what the world needs now more than ever. We can all contribute to this in our own ways. Every action and decision, however small, can make a difference and move us in the right direction.

· *16* ·

Hong Kong Isolation

CHANGE OF PLANS

My initial plan after Dubai had been to fly to India, then travel overland to China via Bangladesh, Myanmar, Thailand, and Vietnam, followed by a two-day boat journey from Shanghai to Osaka, then travel around Japan to conclude my journey. It was precisely at this time that the coronavirus disease (COVID-19) in China had begun to gather momentum quickly, with reports of cases popping up outside Asia. The initial hysteria surrounding the coronavirus was not entirely misplaced, so I decided to radically change the itinerary. As I had just finished my tour of Dubai, I spent the last few evenings there poring over possible routes and combinations.

My plans, like my life, were not under my control. I found the visa process for India both laborious and objectionable. The authorities requested excessive amounts of information, including my religion and other personal data that should have no bearing on a decision whether I was entitled to enter the country. Given that the recent Citizenship Amendment Act passed by India is clearly discriminatory and help-ing to fuel the fire of racial tensions between Hindus and Muslims,[1] it serves as a remarkable distraction from the environmental, economic, and other issues that affect so many people in the country.

I deliberated for a couple of days and decided to abandon that trip because I didn't want the government to use my US$100 to strengthen the societal wedge it had worked so hard to craft. Nor did I want to be in implicit agreement with the application process by completing and submitting the racist immigration form that went against my values. So,

I deleted the data I had submitted and canceled my flight to New Delhi, opting to fly to Hong Kong instead.

After packing my bags on my last night in Dubai, I saw that the U.S. Centers for Disease Control, the gold standard for global epidemiology, had rated Hong Kong as level 1,[2] the lowest of three ratings on the coronavirus alert scale.[3] The border between Hong Kong and mainland China, which was rated as level 3 (maximum),[4] was shut, meaning that it would be impossible to enter even by air, but if I could at least fly to Hong Kong, a special administrative region of China more or less en route to Japan, I might be able to wait it out to see if the border would reopen.

I felt nervous as I boarded the Airbus A380 double-decker plane destined for Hong Kong. I had managed to buy two N95 face masks and a bottle of alcohol gel, which would at least offer me some protection. The atmosphere on the flight was tense. Everyone was looking around, suspicious whether the person next to him might be sick. I tried not to dwell on it and channeled my anxiety into something useful: I wrote copious notes while ideas were still fresh in my mind, such as small details from Dubai and some thoughts about the future.

Upon arrival in Hong Kong, one of the most densely populated cities in the world, and on a Saturday night, I discovered a ghost town. The bus station adjacent to the airport was eerily quiet. I took the bus to Jordan, a neighborhood on the Kowloon peninsula where the streets were nearly empty; the brave few who had ventured out were wearing face masks. The subway was cold and desolate, each rider looking into the other's eyes with a sense of fear and trepidation. A hyper-modern city with high-rise buildings as far as the eye could see, but with almost no one outside, was a scene as surreal as Dali's melting clocks.

My hotel room was tiny, large enough for a bed and a walk-in toilet/shower. Being on the tenth floor, I had a view of the empty street below and other high-rise buildings, which was uninspiring, but at least the room was bright. I went out every day to get some exercise, but I spent most of my time indoors trying to figure out options both in China, in case the border might open, and Japan, where I wanted to go in any case. Hiroshima had an important species of tree I wanted to see (chapter 16).

I kept up to date with events around the world. Before long, the coronavirus had reached Europe, affecting numerous countries, including Italy, which would quickly overtake China's death toll.[5] About a month later, the United States became the world's coronavirus epicenter.[6] As one might expect in such circumstances, people across the globe began to panic. The media quickly shifted all of its attention to real-time bulletins on infection and death rates per country, turning human beings and countries into data points represented in graphs and tables.

I used this unexpected pause in my hectic trip to contemplate what was happening. Billions of people would soon be caught up in the biggest global event since World War II, affecting health systems, social interactions, the labor market, and almost every facet of the way many societies around the world live. Although globalization was the means by which the virus was clearly able to spread so quickly, I began to reflect on humankind's relationship with the environment and the collateral effects, not only on nature, but also on us.

A nationwide study in the United States published by researchers at Harvard University established a link between increased mortality in coronavirus patients and exposure to fine particulate matter ($PM_{2.5}$) small enough to enter deep inside the lung: "A small increase in long-term exposure to $PM_{2.5}$ leads to a large increase in COVID-19 death rate, with the magnitude of increase 20 times that observed for $PM_{2.5}$ and all cause mortality. The study results underscore the importance of continuing to enforce existing air pollution regulations to protect human health both during and after the COVID-19 crisis."[7]

Dr. Vandana Shiva, an expert in environmental conservation, believes that COVID-19 is the symptom of a larger problem; namely, the way in which humankind "violates the integrity of species and spreads new diseases."[8] According to her, "Over the past 50 years, 300 new pathogens have emerged as we destroy habitats and manipulate them for profit."[9] In her article, she cites numerous examples, including the cause of the Ebola virus disease (EVD) outbreaks in Africa that have been linked to deforestation.

Ebola,[10] like COVID-19 itself, is a type of zoonotic disease,[11] meaning that it can be transmitted from animals to humans. The authors of a study that analyzed twenty-seven outbreak sites of Ebola in Africa suggest: "The increased probability of an EVD outbreak occurring in a site is linked to recent deforestation events, and that preventing the loss of forests could reduce the likelihood of future outbreaks."[12] This adds yet another compelling case to the list of reasons why forests are so important, and why our relationship with them needs to be one of humility, respect, and care.

The emergence of COVID-19 had affected the global supply chain in an unprecedented way, and most people, from Argentina to Uganda to Australia, were directly affected by this momentous change. Shortages of hand sanitizer, face masks, and, for some reason that I still don't understand, toilet paper, became common. Even after my return to Berlin a month later, they showed no sign of abatement: many supermarkets had empty shelves. Restrictions of movement began there and in other countries around the world, along with full quarantine in Italy and India, and other countries that would follow suit in an attempt to avoid the complete collapse of their health care infrastructure.

Measures that governments took to coordinate with their health and security services as well as other governments varied greatly. The United States closed its borders to Chinese nationals, followed by Europeans. Germany, France, Denmark, and other countries began to close their borders, all in an attempt to reduce the rate of infection in their populations. In just a matter of weeks, the world looked and felt like a completely different place.

While politicians and central bankers desperately continued to perform CPR on capitalism as it underwent yet another collapse due to its excesses, the poor in the Global North and Global South couldn't observe the #StayAtHome ritual because they had no homes to go to. This included the homeless Americans in Las Vegas who slept in a parking lot even though the city had 147,000 empty hotel rooms,[13] and the innumerable others across the world who don't have a home, putting them at greater risk of infection.

Some businesses asked staff to work from home, others asked them to take voluntary leave; in some cases, people lost their jobs. The airline and tourism industries felt the initial pinch as flights and hotel bookings around the world were canceled. Entire airlines, from budget operators to global players, were grounded. Air travel, predicted to double in less than twenty years,[14] was now facing an almost vertical slope downward. Some CEOs, such as Richard Branson, began to ask for government money.[15] This could, surely, be seen as a supreme irony, given that he is one of a group of businessmen that wants to privatize Britain's National Health Service (NHS),[16] conveying perhaps the ultimate expression of greed: having your cake and eating it, too.

Shares fell in stock markets across the globe, many key currencies such as the euro, sterling, and yen weakened against the dollar, gold prices spiked, and oil prices collapsed, all reflecting nervousness in financial markets due to weakened demand from a guaranteed recession, together with the possible financial contagion caused by a lack of liquidity both in financial markets and in households unable to keep up with payments. These factors caused numerous central banks to slash interest rates, including the United Kingdom's, which lowered its base rate to 0.1 percent,[17] the lowest in its 325-year history.

As a further measure, some central banks took the unprecedented step of applying quantitative easing, essentially creating electronic money out of thin air to buy back government bonds in order to prop up the economy.[18] Such an artificial intervention, which should belong in a fiction book, was implemented to sustain a system that is intrinsically unsustainable and incompatible with the environment. Seen through a pragmatist's lens, this was another fix to keep an imperfect system from collapsing and causing damage to people's livelihoods, but surely it was just a perpetuation of a fairy tale that only truly serves the rich.

On a more positive note, the immediate, coordinated, and forthright response to the coronavirus proved that where there's a will, there's a way. Governments and citizens reacted quickly to an urgent threat and cooperated for their own benefit and for the good of humanity. The Flatten the Curve[19] message swept around the world. From New York to Buenos Aires, London to Johannesburg, to Hiroshima, practically everyone I had met on my trip would soon be at home patiently waiting for the storm to pass.

Some countries, such as Germany, which had based decisions on science and rational thinking, acted decisively to flatten the curve. Angela Merkel's consistently measured, fair, and just behavior helped steer her country to safety quickly. UK Prime Minister Boris Johnson, who wants to privatize one of the country's most important institutions, the NHS, asked retired doctors and nurses, part of the high-risk group due to age, to come back to work. Within a few weeks, twenty thousand staff had returned, showing courage, commitment, and solidarity. In an unexpected turn of events, it was the people Johnson betrayed who rescued him when he fell ill with COVID-19 and had to be admitted to an intensive care unit at St. Thomas's Hospital in London, part of the NHS.

Meanwhile, the United States was in pandemonium. The leader of a country with a population of more than three hundred million people, who has no background in science, told his faithful that chloroquine phosphate could be used to cure the coronavirus. That killed at least one person.[20] President Bolsonaro in Brazil, who is bending over backward to destroy the Amazon rain forest, didn't take the initial stages of the pandemic seriously, just like his counterpart in the United States. Mass graves would be dug in two cities I had visited earlier on the trip—Manaus in Brazil[21] and New York City[22]—clearly reflecting two racist, misogynistic leaders who do not respect science, nature, or humanity.

After spending several days almost exclusively confined to my hotel room in Hong Kong, interrupted only by my daily walk and visiting the local 7-Eleven to buy food and water (where, strangely, I was offered some hashish outside), I carefully analyzed my options. China was still on high alert and it didn't look like the border, just a few hours from my hotel, would open anytime soon. More borders would be shut, so rather than capitulate and book a flight home to Berlin, I decided to roll the dice one last time and fly to Japan. Even though it had been upgraded to level 2 by the CDC,[23] at least I'd be allowed to enter the country.

Hong Kong International Airport looked emptier than when I had arrived. Almost all of the shops were shut; the eyes of the staff, peeping above face masks, looked sad. Many flights had been cancelled, and only a small group of people could be seen eating lunch in one of the few eateries still open. Eleven other passengers and I boarded the flight to Tokyo. The nervous cabin crew handed us Japanese immigration forms together with an additional questionnaire from the Japanese government related to the coronavirus, which requested specific information ranging from flight details and seat number, where I had been, where I was going, and additional contact details, all to help with traceability should the outbreak intensify in Japan. Against all odds, the final leg of my journey had begun.

· *17* ·

From Hiroshima with Love

BULLET TRAIN

I was traveling at two hundred miles per hour in a Japanese bullet train with Wi-Fi and air-conditioning, and I ate some matcha (green tea) ice cream as a treat. The train calmly moved at high speed in near silence as I oscillated between writing notes and meditating on the rivers and numerous forests that zoomed past with a million shades of green, from the palest to the lushest I have ever seen. Large stretches of nature were punctuated by densely populated areas and factories as I made my way from Tokyo to Hiroshima. At one point I saw a wind farm in the distance, reminding me of my time in Denmark and the potential for the Earth to provide us with clean energy that can readily be harnessed.

Back in Africa, I often traveled on overcrowded buses, clinging to anything I could as the driver overtook on a blind bend or dueled oncoming traffic, something that happened frequently in Latin America as well. I felt secure in Japan, safe to walk around with my camera at night, something unimaginable in almost any country in the Global South that I had visited. Neither did I worry about food or water hygiene in Japan. All of these comforts and supposed securities would, however, prove to be supported by the same illusions I had seen earlier in Dubai, London, Miami, and elsewhere.

The train made a short stop at Kyoto, where I reflected on the magnanimous climate protection treaty drawn up in 1992 and signed by 192 parties,[1] and questioned why so little had been done since most of the cities in the Global North I had visited were major polluters.

Apathy is abundant. Even when people know about the magnitude of the problems, some attitudes might be perceived as flippant, such as that expressed by a friend of mine who works at a leading environmental organization. She complained about the cancellation of the UN Climate Change Conference (COP 25) meeting in Chile,[2] not because of protests that left twenty people dead but, rather, because she wouldn't be able to enjoy the warm and sunny weather during the southern hemisphere's spring as the event's new location would be in Madrid in northern Europe at the end of autumn, when it would likely be dark and cold.

Even when the wheels are falling off, some "climate experts" are more interested in their own entertainment than the well-being of the planet because they have spent too much time in meetings, drafting proposals, and climbing the slippery ladder for prizes and prestige in academia, government, the public domain, or anywhere they can find plaudits to satisfy their narcissism. The real experts are those who have connected mind, body, and soul like a perfect equilateral triangle—at their Cartesian center is their appetite to serve others and inspire hope. Of course, some are endowed with advanced degrees, whereas others are exclusively educated in the world's most comprehensive library: the field.

Humanity's universal place of worship isn't restricted to the shrine, temple, church, or other man-made structure of their choice; rather, it is contained within every natural field, forest, lake, and river where the blueprints of the most comprehensive and diverse compendium of DNA live. Although knowledge itself needn't be venerated, it should at least be respected and an attempt made to understand and preserve it. Nature's library includes both all of the life that we can see, such as birds and insects, and that which we cannot see with the naked eye, such as the intricate network of fungi networks beneath forests (chapter 3) and the cells of all living things that exchange nutrients that keep the engine room of life running.

My reverie came to an end when the train pulled into Hiroshima station at 1:55 p.m. on Thursday, February 27, 2020. I only had a few more photos to take, and one final interview to secure. It was at that point that I could see the summit of my own journey. I felt pleasure and joy, with a small measure of trepidation.

BLACK RAIN

"We were walking along the hillside road, when all of a sudden a very bright light flashed. Next came an overwhelming blast of wind with [a] big sound. We were almost blown away. Promptly, my sister covered me with her body, and we lay face down on the road."[3] Soh Horie was nearly five years old when the atomic bomb detonated in Hiroshima. Had it happened a few minutes later, both he and his sister would have been on a nearby hill and may have died from the flash of light and the blast.

"I still remember two persons. One was a junior high school student. He was burned all over his face, and his nostrils were clogged by peeled skin. He was breathing painfully through his mouth."[4] Mr. Horie's mother removed the skin from the young boy's nose. "The other was a young girl. She was burned and the patterns of her dress were printed on her arms."

Mr. Horie was playing near his house when black rain started to fall. Only later did he realize that the precipitation was, in fact, fallout. His father's underwear was among the laundry that had been hanging out to dry, and it turned black. His mother tried to wash out the color but couldn't because the black stains contained radiation, so they decided to donate it to the Peace Memorial Museum located close to the hypocenter.[5]

Soon after the blast, a local Tenrikyo (a sect of Shintoism) preacher brought an object of worship to Mr. Horie's home and began praying, "but he was unable to save anybody."[6] That evening, Mr. Horie and his other sister were taken to their aunt's villa. He recalled the sky being bright red from the houses still on fire. One of the most indelible memories for him was the smell of cadavers every day for a month. He asked me to imagine the smell of countless decaying corpses.

During the interview, Mr. Horie referred to his computer, where he had stored lots of maps and photographs. He showed me his family photo, taken in 1942, where he was sitting on his mother's lap, his father dressed in his naval officer's uniform, and his siblings. Then he showed me a list of the types of cancers that his family members died from. His father passed away six days after the attack from an acute illness. Mercifully, Mr. Horie won his battle with a lymphatic malignancy thirteen years ago.

Today, he remains active through talks he gives to the general public about his experiences, the courses he gives to children on how to plant rice fields, and his antinuclear activism about which he is both anxious and passionate. He showed me a map of the Ikata nuclear power plant about sixty miles away. It sits on the Japan median tectonic line, causing him great concern about the ripple effects an earthquake would have given the nuclear accident in Fukushima that followed a tsunami in 2011. He reflected a worry that I, too, have about nuclear power.

Nuclear fission was once seen as the panacea for the world's energy problems, promising a limitless future that could drive economies forward and remove limits to what humankind could achieve. Even today, many countries rely on this technology, which is fraught with risks, as evidenced in Fukushima and Chernobyl. Alternatives, such as wind (chapter 8) and solar (chapters 2 and 7), could dramatically reduce the need to use nuclear and hydrocarbons as fuel sources, so I was puzzled how and why such a modern country could still be so attached to such an archaic, dangerous, and detrimental technology.

Then Mr. Horie proudly told me about Japanese inventions, including the recent Nobel Prize winner Akira Yoshino's lithium ion battery, subsequently describing the country's strong industry, includ-

Soh Horie showing his family portrait taken in 1942. Photograph by author, 2020

ing that about one-sixth of the world's cars were, at the time, made in Japan. I was there to listen and learn from him, but it was somewhat painful to hear him extolling the virtues of the Japanese car industry when I knew where some of the metals came from and what the speculative exploration work entailed. It will inevitably dredge the ocean floor to develop an industry that is not only unsustainable in its current form, but plain dangerous. The final part of his presentation was a broad view of chemicals, including depleted uranium-235, which he said was used in both Iraq wars, and his antiwar stance related to his antinuclear activism: "No more war; it's destroying the planet, and I believe that it [his activism] is my final job."[7]

It was an immense privilege to have met Mr. Horie, an experience I will treasure forever. He had arrived early to our meeting, impeccably dressed in a pressed shirt buttoned up to the collar, thick sweater and coat, and he wore a kind smile. As a parting gift, he gave me a pencil made of a small branch with the bark still intact. My face lit up, then I quickly frowned because I didn't have anything to give him. I told him and Mark McPhillips, who had assisted with the translation, and Erika Abiko, the owner of the Hachidori-sha Social Book Café,[8] who had helped organize the interview and host us that evening, that I would leave a small gift for each of them with Erika before I left Hiroshima. This at least bought me some time to find something to say thank you.

SIX TREES

The other key reason I wanted to travel to Hiroshima was to visit six fabled ginkgo (*Ginkgo biloba*) trees that had survived the atomic bomb. Each of the surviving trees stands about one mile from the hypocenter, forming a loop around it. I hired a bike at the hostel and began my pilgrimage, something I had imagined for months prior to my arrival. As I cycled across bridges and along the river banks, life passed me by slowly because I myself had slowed down.

At my first stop, Myojoin Temple, an extraordinarily beautiful structure with traditional roof and wooden features, I was greeted by purple and white cabbages that had been arranged as ornaments, creating purple and white towers that would give exuberant flowers come spring. Then I saw the first ginkgo tree.

Set inside a lovely garden within the temple's grounds beside a Buddhist shrine, the ginkgo tree, surrounded by other trees, stood about thirty feet high, towering stoically and devoid of leaves because it was still winter, with moss growing along large stretches of its trunk. If Salgado's tree nursery in Brazil (chapter 3) was a neonatal ward, then the ginkgo trees were humanity's veteran heroes and heroines—figures who had witnessed suffering, withstood violence, contemplated time, given shelter to wildlife, and released oxygen for well beyond living memory.

As I walked around the temple grounds, I discovered more tree survivors: the Japanese sago palm (*Cycas revoluta*) had several thick, dark trunks ranging in height up to twelve feet, with crowns similar to palm trees; then I saw the Japanese black pine (*Pinus thunbergii*), which looked as if it had been watered from the fountain of eternal youth given the intense green colors of its leaves and nothing that indicated to me that it had had any traumas. Each of the surviving trees I saw in these grounds had a laminated information sheet draped like a necklace around its trunk, including its common and Latin names and a statement that it had survived.

My next stop was Anraku-ji Temple, where the surviving tree stands next to the main entrance. The tree was so tall that I had to walk to the end of the parking lot in front of the temple to frame the whole tree in the shot. From that vantage point, I could see that one of the branches was so big that a hole had been made in the roof's entrance so the branch could pass through it.

After crossing the Misasa Bridge in the north of the city, I took a break to sit by the riverbank. As it was a Sunday, lots of families were walking, people jogging, and many cycling. I even saw a man riding a thirty-six-inch unicycle, which brought a smile to my face and others as he sped past. This, together with the innumerable cycle lanes, bike shops, and other cyclists in Hiroshima, made it feel more tranquil and peaceful than Tokyo and drew me closer to the city.

I had been to other parts of the world that had faced tragedy both on this and on previous trips. What struck me here was the calm, genuine warmth I felt from everyone I met, even though the trauma of the atomic bomb was imbued in everyone's souls. Seeing those trees didn't necessarily make me feel different about the climate science I had read; rather, it made me realize how simultaneously fragile and strong nature is. She has many mysteries, and I was in a quandary as to how those trees could have withstood such high temperatures and the nuclear fall-out that came after the bomb had been detonated.

Life itself resists death, but when it succumbs, it simply cannot overcome the forces that act against it. I recalled the bees in Europe and Africa (chapters 10 and 11), which, unfortunately, I would not see in Japan because it wasn't warm enough. It is a sad thought, however, that as global temperatures rise there may well be a March first in Hiroshima hot enough for bees to be flying around.

The third stop was Hosen-ji, a modern temple made of marble and concrete without a gate, making views of the stunning tree accessible to passersby. The two staircases leading to the temple's main entrance form two arcs around the ginkgo tree, placing life at the center of a religious building. Due to the coronavirus outbreak, I couldn't enter that temple or others, but even standing outside and walking up and down the staircase, it felt as if the tree had been infused with the building. Indeed, it was highly probable that the tree's root system extended far beneath the base of the building's foundations and was somehow connected to the graveyard that formed a U shape around the sides and back of the building.

I cycled for another couple of miles and reached my fourth stop, Josei'ji Temple. Because the map I downloaded had mistakenly directed me to the back entrance, I needed to traverse the temple's graveyard to get a view of the tree, meaning that in order to see life, I had to walk through death. The two diametrically opposed states coexisted in that temple and reflected the hope and fear I have about the destiny of humankind, nature, and the planet.

The penultimate stop was the tree at Senda Elementary School, which was shut because it was a Sunday. However, it's impossible for me to be certain whether I saw part of the actual ginkgo tree from the street because so many other trees were in the way.

My final stop was the Shukkeien Garden, an impeccably arranged haven set around a lake. The attention to detail in the garden, from the plants and flowers meticulously pruned and arranged, to the overall way it was curated, probably explains the awe I have for Japanese art. As I walked around the grounds, I did the honorable thing: I accepted two young ladies' requests to photograph them, using their cellphones, in their traditional kimonos. It was as if I had stepped onto a film set, because it would have been impossible from the composition of the photo that I had taken of them to determine what year it was. It could have been a few years, decades, or even longer before then.

It was difficult to see the ginkgo tree at first because it was surrounded by so many other trees, but finally I figured out which one it was. The final tree on my quest was possibly the tallest of them all, leaning at a thirty-degree angle, able to do so because of the incredibly strong root system beneath my feet. After taking the obligatory photos of the tree from different angles, I sat down far away enough to see it in its glory and entirety. I observed and breathed, nothing more. My journey was nearly complete.

I reflected that evening on the significance of what I had seen. Why did those ginkgo trees survive and others did not? Why was Mr. Horie so fit and strong despite all of the traumas he had witnessed and the physical suffering he had endured? One answer might simply be resilience—the idea that no matter where you put a living species, its drive to live and strive will overcome all odds. In this sense, I had seen several miracles.

Before my departure, I visited Erika's café for a final time and brought prints of a series of poses of a European honeybee that had been playing hide and seek with me while it walked along the length of a leaf. I earmarked the best pose for Mr. Horie, and let Erika choose one of the remaining two for herself, asking her to pass on the last photo to Mark. We shared a warm hug and wished each other well.

As I walked through the cool streets that evening, everything seemed fine. The locals were living life as usual, calmly, stoically, and quietly. I crossed the Honkawa Bridge and walked through memorial park one last time. A couple of cyclists overtook me on the narrow sidewalk, which was dark and wet. I remember feeling a cool gust of wind against my chest.

Part V

JOURNEY'S END

• 18 •

Lessons Learned

A BLESSING IN DISGUISE?

$\mathcal{U}$pon my return home to Berlin in March 2020, I had time to unpack my bag, review my notes, interviews, and photographs, then take a long, hard look at myself, my life, and what had happened since I began the project in May 2014. After a brief realization that the journey was over, I quickly resumed my routine of visiting the American Memorial Library in Berlin to work on my manuscript in the airy, spacious, and well-lit surroundings. A few days later, the quarantine measures quickly put in place due to the coronavirus meant that I would spend the next few months working at home in isolation.

The coronavirus pandemic eclipsed the Zika crisis I had navigated in the Amazon in 2016. Public fear rose to the surface once again, aided by a media frenzy. People began to ask when things would go back to normal. However, if "normal" meant a collision course with climate catastrophe, then perhaps the coronavirus provided an opportune pause to the senselessness of unbridled consumption. Why did the media not focus on some of the positive things, such as the reduction in carbon emissions because people were flying and driving less?

Nature, as I saw in the Atlantic forest in Brazil, has a remarkable capacity to heal when the aggression against it is suspended. About a year after my stomach surgery, another study of my esophagus was done, and it was found to be back to normal. By creating a new valve between the stomach and esophagus—and, with it, a new barrier—my body was able to heal itself from stomach acid that shouldn't have traveled upward. I remembered this when, a few months into the coronavirus pandemic, I reviewed satellite images of nitrogen dioxide (NO_2) levels, which were down in European cities.[1] Mount Everest was visible from more than one hundred miles away.[2] Dolphins could be seen in the Bosphorus.[3] More bike lanes were opened up in Bogotá, Philadelphia, and Berlin to promote cycling.[4]

The drastic reduction in flights reduced pollution high up in the sky an undetermined amount, as did less road traffic to and from airports, and reduced oil extraction, itself highly polluting. The southern jet stream seemed to be healing, thanks to a collective action to reduce the use of chemicals that harm the ozone layer through the Montreal Protocol,[5] which showed that concrete results happen when efforts are coordinated.

At times, I wondered whether the coronavirus pandemic was a dress rehearsal for something bigger, such as, for example, if global warming and population growth were to continue at current projections. One of the biggest problems that humanity faces with regard to climate change is cognitive dissonance: in spite of incontrovertible evidence indicating that we are fast approaching overcooking the planet and, with it, dire consequences, people choose to ignore this and continue a dangerous cycle of greed, consumption, and waste.

During the first few months of the pandemic, out-of-control consumerism went cold turkey. Two risks could be foreseen. The first was that life around the world would go back to pre-coronavirus times if, or when, a vaccine or other appropriate arrangements could be made. That is, we would return to what some might call "normal," a term that I would avoid because that normality, in fact, means overusing the planet's resources and resuming a flawed system that will continue to exacerbate inequalities, iniquities, and injustices while accelerating the speed at which we will breach the tipping point that is already fast approaching, if indeed it has not already been breached.

The second risk I perceived was that once normality returned, then much, if not all, of what people would have learned during the difficult period might be forgotten. What the first few months of the quarantine period in Asia and Europe showed me, because I experienced both, was that people can live with less—that it is possible to live a simpler, more fulfilling life without having to rely as much on the material world.

On May 6, with more than seventy thousand deaths in the United States, a number far higher than the number of American troops who died in Vietnam,[6] President Trump declared that the country would be open for business very soon, commenting, "Will some people be affected badly? Yes. But we have to get our country open, and we have to get it open soon."[7] Money was more important than lives, trade more important than equality. My only hope is that the new normal will be far more harmonious with the environment and other people than the old normal and that humanity can show a deeper commitment to reversing the current trajectory.

CLIMATE ENGINEERING AND TEMPERATURE

Notwithstanding divine intervention or another major event orders of magnitude larger in impact than the coronavirus, any reasonable and sensible person could conclude that global warming is here to stay for generations to come: cherry blossom trees had started to flower in Hiroshima even though it was still winter; cities from Berlin to Tokyo were unseasonably warm during my journey; the oceans were warming and acidifying; Antarctica broke the 20° Celsius record in February 2020.[8] Such signs were ominous and indicated that more radical approaches may need to emerge sooner than anticipated.

Climate engineering is an, as yet, unproven discipline that seeks to implement large-scale engineering projects that can cause a significant effect on weather dynamics. Such projects might include spraying sulfuric acid high up in the atmosphere[9] to mix with water and so create aerosols to deflect sunlight and cool the Earth; dumping iron filings into the ocean, which will cause algae to grow and absorb[10] CO_2, bringing temperatures down; injecting a chemical (e.g., bismuth triiodide) in a specific part of the atmosphere that would cause "cirrus cloud thinning,"[11] which would result in more radiation being reflected and the Earth cooling. These are just some ideas that may come to fruition, even though their side effects cannot be accurately predicted given the scale and large number of variables involved.

All of this would be fine if it were a thought experiment or a mathematical model done at a school or university, but some people are quite anxious to start tampering with the divine beauty of the natural world—itself a rape of the environment and a sure sign of arrogance and ignorance. In 2012, businessman Russ George is alleged to have dumped one hundred tons of iron sulphate in the Pacific Ocean off the coast of Canada[12] with the aim of mitigating some CO_2, but it will be very difficult to measure the collateral effects of this deed.

Rather than try to preserve the natural world in its current state and look for ways to reduce emissions and sequester carbon, ostensibly, some who work in industry, politics, and elsewhere will try to exploit the desperate situation that the planet faces to justify these actions, or simply for the thrill of the intellectual and grandiose pursuit of trying to play God. A recent paper published by leading academics who seem to be proponents of geoengineering, makes their position clear: "Relatedly, the concept of the Anthropocene, with its emphasis on the planetary impact of human activities, may further normalize climate geoengineering technologies as potential tools for conscious planetary management."[13]

In the age of relativism, normalization is king. Buying stolen jewelry in the form of a gold watch mined in deplorable conditions that has been exported clandestinely out of the country, filling up the tank with gasoline that is heavy in blood, and eating a hamburger from a butchered forest have all become normalized to the point that they are actually taxed and a citizen could be punished by law for stealing any of these things, even though they are essentially already stolen goods.

To use the planet as a large laboratory for megalomaniacal ideas with unpredictable results is not only dangerous to the natural world we depend on, but a blatant disrespect to what's precious: the planet. The deep-sea mining operations discussed toward the end of my journey in Africa, for example, are part of an invisible effort to keep the capitalism show on the road.

Meanwhile, the global average temperature, which is changing quickly with respect to time, is positively correlated to carbon emissions. As much as I would like to shy away from this awareness, we simply haven't much time left. We are no longer in the eleventh hour. We are in the eleventh hour and fifty-ninth minute. As we draw nearer to the infamous 1.5°C tipping point,[14] which may happen by 2030, if not sooner, I have come to realize that this might be the final opportunity for reflection and change before the real chaos becomes systemic. This will happen once positive feedback mechanisms (i.e., a snowball effect) begin, and cook humankind like the mussels boiled in their shells as they lay on a beach in California in summer 2019, an image that has stayed with me since I first learned about it (chapter 11).

Given the scale and magnitude of this crisis, and the fact that the temperature anomalies are continuing their relentless upward trajectory, the natural and logical inference is that the planet will get hotter. Much hotter. The scientific literature points to this, and it makes more sense to understand this as being part of the brave new future rather than to become vegan, stop flying, meditate every day, close one's eyes, and hope that the problem will disappear. Because—and it is with a heavy heart that I write this—it seems that the future will simply be less bio-diverse and problematically warmer.

A survey of some key crises in the Global North that have happened since the turn of the century reveals predictable responses by governments. September 11 engendered a war on terror resulting in the killing of hundreds of thousands of innocent civilians in countries that were unconnected with the attacks. The 2008 financial crisis caused by greedy bankers who overvalued high-risk mortgages and played musical chairs, causing the financial meltdown, was handled by a government bailout.

Meanwhile, environmental disasters are handled in a different manner altogether. The Fridays for Future initiative led by Greta Thunberg, with millions of people a month protesting each Friday because the world is on fire, led to . . . nothing. The Amazon was on fire in 2019—also nothing. During the 2019–2020 Australian bushfires, in which one billion animals died and 15.6 million acres,[15] roughly the size of West Virginia, were burned to a crisp, the Australian prime minister went on holiday to Hawaii[16] and the coal industry continued—so essentially, nothing. The 2003 summer heat wave in Europe killed 35,000 people[17]—nothing; in 2015, a heatwave in India killed 2,330 people[18]—nothing; and yet another major European heat wave in 2019 killed 1,500 people in France alone[19]—nothing.

In spite of these clearly dire events, there wasn't a shutdown of airports, a ban, or even a modest rationing of fuels. The European Commission pledged one trillion euros over a ten-year period to tackle climate change and to make Europe net carbon neutral by 2050.[20] Per capita, the pledge represents less than 220 euros (US$240) per year. This, together with the time line, considering that current projections suggest that humanity will overshoot the so-called tipping point in 2030 unless something radical is done before then, clearly indicates that both the budget and proposed time frames are completely out of touch with reality.

Humanity is on the *Titanic*, and it's crashing around us; yet, the music is still playing, people are smoking cigars and having a whale of a time while others are drowning in ice-cold water in the middle of the night. Businesses and governments behave as if we have no climate crisis and press forward quite literally as if nothing's wrong. For their part, the Global North masses follow obediently because they have been distracted by less important issues and feel entitled to continue their high standard of living even if it comes at the expense of the environment and other people.

This set of behaviors is a blend of cognitive dissonance, denial complex, and outright cynicism supported by violence and psychosis, in the form of violations of human rights and ecosystems. That it is normalized perhaps indicates some level of negation, but the underlying symptoms of this malevolent disease are greed, entitlement, and ignorance.

That custodians of power and captains of industry consciously choose to behave in this way is unsurprising, given that they are a small group of people who wield a lot of power, many of whom had it bequeathed to them through nepotism and so-called democracy, which is perpetuating the neoliberal agenda of expansionism, aggression, and humiliation. As we have seen during the coronavirus crisis, climate change could be significantly reduced tomorrow, but we have no desire to do so even if the end result is living on half a livable planet that is 5 or 10°C hotter because profits are too important and we are unwilling to abruptly change the current way of living, which is absolutely unsustainable.

A clear chasm is beginning to emerge. Individuals are beginning to speak up and have their voices heard, even if that comes at the cost of their lives, such as the brave defenders of the environment in Latin America, a notorious hot spot on this small planet. Some young people, led by Greta Thunberg and numerous other youth activists, such as Ugandan Vanessa Nakate, together with more seasoned professionals, such as Kisilu Musya, are fighting for a better future for everyone and everything on the planet.

An important consideration when discussing carbon emissions is the fact that, for all intents and purposes, carbon is cumulative, meaning that even if we have a modest reduction in carbon emissions, we would still have more carbon in the atmosphere than if we completely stop emissions, which ultimately means a hotter planet. Thus, the malevolent loop of consumption, destruction, and recklessness needs to stop in order for the Earth to cool down, both literally and metaphorically. And it's actually very simple. First, we need to reduce unnecessary consumption and waste. People value and throw things away unnecessarily. From the Michelin-starred restaurant in Copenhagen to the music school in Asunción, clear evidence shows that such a reduction is, indeed, possible on an everyday level, as these ventures use their creativity to innovate and reduce their carbon footprint.

Another potential solution is planting more trees, which itself leads us to a quagmire. In 2015, the world had an estimated three trillion trees.[21] But trees are not a zero-sum game. The trees that burned in Australia and Siberia, or the trees that are being felled in the Amazon and the Naha jungle in Mexico, cannot be easily replaced, if at all. Not only do these trees release important carbon when they are destroyed, but the ecology that collapses when they are gone has other ramifications beyond just carbon release. Moreover, the new trees don't have the capacity to retain as much carbon early on because they are too small.

On the other hand, Salgado's forest is a good example of what individuals can achieve through time, dedication, love, and education. Preservation of and adoration for the natural world are just as important as the elevated consciousness we need in order to make more informed personal choices about what we consume.

NEW ALLIANCES AND PARADIGMS

The trip to Hiroshima, where I visited the surviving ginkgo trees and heard Horie's firsthand testimony of his pain and suffering following the detonation of the atomic bomb in 1945, revealed how much the world had changed since World War II. Hiroshima, Nagasaki, Pearl Harbor, and the concentration camps in Germany and Poland. These, alongside other traumas that involved large numbers of deaths, would be consigned to history books, replaced by new alliances and partnerships that would emerge between unlikely partners bound together by commerce. The only limit was the horizon, and nobody would be able to stand in their way. Then China came along and began to grow at breakneck speed, eclipsing many economies at the expense of the environment.

For many countries in the Global North, war has mutated into games, marking a continuation of the entitlement this region conferred on itself. War has been normalized and accepted in society through the proliferation and adoption of violent and ever more realistic role-playing war games, encouraging young people to learn the art of war from their own homes. Violence in Hollywood films also helps to condition the masses to accept excessive and uncontrolled use of force against innocent people. The cherry on the cake is impunity: instigators of war do not face trial, creating yet another incongruence in the values of the Global North.

This large-scale desensitization, together with Bauman's idea of liquid risk and the resultant fear, explain how the public has been misled and conditioned into thinking about random and improbable exogenous risks and developing a narrow-minded bunker mentality, as evidenced by the increasing number of people in the Global North who elect openly xenophobic and racist leaders. Hell isn't a place where one goes after death, having committed sins on Earth. It exists on Earth, and I saw it next to the coal mine in Colombia, in the burned-out homes in Argentina, and in the southwestern townships designed for black people during apartheid that still stand today.

The environment is perceived to be an inanimate box from which powerful nations can extend their arms and take what they can. Even when they have too much, they want more. But whether the resources are on one's own soil or otherwise, they belong to the Earth. As humankind, we are merely custodians and guardians of the resources, rather than their outright owners. The coronavirus pandemic has demonstrated how inextricably linked our present is, and how intercalated and fragile our present and futures are.

Meanwhile, in almost every country I have ever visited in the Global South, most people are struggling to survive. They often live in quite precarious conditions, meaning that not only do they produce less carbon emissions, but they fight to get through the day and many of them are ill-equipped to face the challenges ahead. The theft of their resources and the creation of a woeful environment have essentially trapped many in a dangerous situation.

Our relationship with the environment is violent, ignorant, selfish, and ruthless. The forces at work to perpetuate aggression toward the Earth do not seem to show any signs of abating; rather, pressure to increase shareholder returns seems to be the main driver: profit over the Earth and the minions. The smokescreens are the carbon-neutral initiatives set to take place by 2030, 2040, and 2050, which provide some respite to the uninformed but sensitive folk in the Global North and Global South that something is being done about increasing temperatures, weather instability, deforestation, inequalities, and other issues, all of which go hand in hand.

Indeed, a lot of innovation is happening, especially in the Global South, where extreme conditions have encouraged the exploration of new ideas. The Global North is, generally speaking, conservative and old school. What are its engineers thinking? Have gasoline, cobalt, and lithium, which continue to be used in their innovations, suddenly become ecologically friendly chemicals from ethical sources?

This laissez-faire approach to the environment cannot be attributed solely to the ruling elite of Europe and the United States. Many people in the Global North, including myself, live beyond our means. Living an extravagant lifestyle beyond the means the planet can provide is a personal choice. That being said, many people, both those honored in this book and many others, have dedicated themselves to living within their threshold, which, put simply, is three tons of CO_2e (carbon dioxide equivalent)[22] per annum. If people are unwilling and unable to contribute toward living in a more sustainable way, the probability of humankind having an unhappy ending will increase.

The excessive use of oil products, from coal and jet fuel to superfluous packaging, to drive economies forward means we are at a critical point of fossil fuel exploitation. India and China are the major coal consumers and are well aware that coal drives their economies, as well as others in the Global North such as Australia, Poland, and Germany, who have large deposits of coal. The latter see themselves at being put at an economic disadvantage were they to stop mining it, but this isn't necessarily true.

Governments want to be reelected, and shutting down a coal industry doesn't seem to be on the horizon even when a natural disaster, such as the forest fires in Australia in late 2019/early 2020, rears its head—a direct result of climate change. Surely the thirty thousand coal miners in Australia[23] could be retrained in another industry, such as renewable energy. Canada is set to exploit the country's tar sands, one of the worst polluting sources of fossil fuels, and oil-producing countries in the Middle East and Africa show no signs of abatement. It's about money, jobs, contracts. The environment is an invisible and tangential service.

Interestingly, the petrochemical company BP, which has operations in seventy-nine countries around the world,[24] announced that it would be net zero carbon by 2050.[25] This, at face value, sounds like an act of benevolence, but the set date is actually twenty years after we may have crossed the temperature tipping point, or thirty years too late because all of the science tells us that the oil should stay in the ground. Moreover, being net carbon zero doesn't mean that there aren't any negative impacts in offsetting emissions. This company is responsible for the Deepwater Horizon disaster, one of the biggest ever oil spills in history, in which 4.9 million barrels of oil leaked into the Gulf of Mexico.[26] That same year, BP paid its executives generous bonuses.[27] Today, it continues to knowingly and happily assist countries around the world to pollute even more.

The reason why we have panic about the coronavirus in the Global North and almost no fear about climate change is because governments' plans for the next few decades in Europe, the United States, and Japan remain largely unchanged. Cars will continue to be built using foreign material that will burn fossil fuels, energy will continue to revolve around non-renewables, and agriculture will be leveraged using more GMOs, agrochemicals, and reliance on foreign supplies.

Coal and all other fossil fuels are doomed because they are finite, hazardous to human health, and disastrous for the environment. Humankind must stop using these chemicals immediately through divestment, retraining, and creating new infrastructures, all of which could provide new jobs and a future for all, rather than the myopic, dimwitted, disastrous ending that the current trajectory suggests.

Jumping up and down screaming, "The house is on fire! The house is on fire!" will do no one any good. As a wise nun in Tanzania, Sister Martha, once told me, "People don't learn from facts; people learn from people."[28] It was part of a conversation I had with her in 2013 that helped me focus on people's own stories when contemplating my upcoming journey, rather than write a unidimensional book based on facts alone.

With so many distractions in the form of fashion, the media, drugs (whether illicit or prescription), alcohol, and tobacco, it's easy to understand how it's even possible to be desensitized to others' misfortunes. But, now, humanity is in the midst of the Anthropocene and passing through uncharted water. Governments in the Global North have used their expertise and tradition of issuing more debt to get themselves out of trouble and keep their economies alive, but government bonds and bailouts cannot bring insects back to life or plant a trillion trees overnight.

Our real debt is to the environment. Carbon debt, water deficit in farms, rising sea levels, pollutants, and other collateral effects have largely arisen due to humanity. Our debt is also to those who live in misery to provide the comforts that we take for granted. What is sustainable opens up a quagmire, because we have no clear-cut answer. In the end, we will always be indebted to the environment as long as we take more than we give. Another fallacy I discovered during my journey was about green transport, be it biofuels or electric cars: while it reduced the overall carbon footprint, it created new problems, namely, environmental destruction to procure the raw materials that make the solutions purportedly green.

Ultimately, what's required is a paradigm shift even more dramatic than the opportunity the coronavirus presented us to reevaluate what we are doing. The notion of what is green and what isn't has to change, along with some basic understanding of science. Although achieving modest reductions in carbon is always welcome news, what we really need is carbon sequestration; however, this is more likely to require massive financial investment, legislation, and enforcement to be truly effective. In the meantime, everyone can be involved in enacting positive change on some level, either by planting trees, eating local food, not wasting, or any other action that cuts an individual's carbon footprint.

Knowledge, wisdom, justice, politics, democracy, economics, ecology, and other elements will all play their respective parts, like instruments in an orchestra to create the sound of a symphony. But it is the auditorium—the planet itself—that will ultimately decide how much longer it can continue to play in such an asymmetrical way in which key voices have been drowned out and suppressed.

Bravery comes with empathy. Like Tatiana and Karen in Colombia and Argentina, respectively, who accompanied and supported people even though it came at great personal risk, possibly including their lives.

Like Simon and Kisilu who care for nature and found ways to make an enduring impact, both within their communities and beyond. Like the hundreds of journalists who collaborated on the Panama Papers, sick of the hypocrisy and injustice of a clandestine offshore tax system that only serves the rich and deepens the divide between rich and poor.

Any hope I have for the future is rooted in the hearts and minds of the great people I met during my journey. They are making positive changes directly through their work, and they have inspired me and those around them to do the same, which I hope continues to propagate.

———— ✑ ————

This book is the second part of a trilogy. If I look at most of the musicians whom I interviewed in my first book, *Music and Coexistence*, many of their lives are worse today from the point of human rights: less freedom of expression, more racism and xenophobia, with several having left their countries of origin looking for a better life. The resulting brain drain contributes to the entropy and uncertainty that exist in this brave new world.

My next book will be focused on bees because they are the single most important living organism on the planet. Bees are being eliminated at an unprecedented scale that is impossible to imagine, so it is in their honor that I have already begun to prepare that manuscript. If I can help to save just one of the estimated twenty thousand species, then I feel I will have made a positive contribution.

Daniel Silva, my dear friend who filmed me in Kenya and Uganda, spent six months editing the film titled *My Little Drop of Honey*, which is the translation of my first name. It took him longer than expected to edit the film because he wasn't sure what the real story was. Was it about writing this manuscript, the creative process of being a writer, the struggles of finding an agent/publisher in a saturated book market, bees? The first cut was a transverse view of all of those issues.

Eventually, Daniel decided on the secret center of the film, and he recently sent me the final cut. We had a lengthy discussion about some specific scenes and the ending. I agreed with his assertion that the reason I feel connected to bees and would prefer humanity to be like them is because I am deeply disappointed with humankind in general. That being said, I have not lost faith in humanity.

The reason I continued writing this book to the point of completion was to honor the people whom I had met on the road: Kisilu, Simon, Mr. Horie, Karen and the farmers in Argentina, Tatiana and the community she supported in Colombia, professor Hansen in Denmark, and the other interviewees, all of whom were genuine and showed me generosity of spirit even when some of their lives had been turned upside down. Everyone I met shared pearls of wisdom that have helped me to find strength I previously didn't know existed. I am stronger because of their strength. I am less unwise because of their wisdom.

Afterword

$\mathscr{H}$ope could be characterized by strength, fortitude, and resoluteness. Perhaps hope is the desire and wish for things to be positively different. In these senses, I have hope. But I think that it's almost impossible for me to talk about an excess of personal hope for the future having completed the trip and digested my notes, photographs, interview recordings, and film footage. This book was partly a labor of love and mostly the result of a journey that I had to complete because it was the right thing to do.

My heart, eyes, and soul changed with respect to time in a kind of metamorphosis, from an analytical person who only saw the world through the lenses of fear of the future and optimism, to someone who feels closer to the environment and others and is willing to accept whatever will come because, in the end, I can only play my part. My only wish is to make a contribution to the world that will have a positive influence.

Meditation helped me during some difficult moments, such as being in isolation after I returned to Berlin due to the coronavirus. I have meditated several times a week since 2017 using a technique that has two components. The first is a set of breathing exercises that lasts for thirty minutes, helping to regulate my breathing, bring me to a calm state, and reduce the excessive thoughts that so often fill my mind. This is followed by an analytical meditation involving a series of exercises, the most important of which for me is to list between five and ten things I am grateful for and reflect on why I have chosen them. The list changes from day to day, but it's usually as follows:

1. My friends
2. A roof over my head
3. Food in the fridge
4. The sun
5. The rain
6. The Earth
7. Air
8. All living species
9. My freedom

Gratitude is essential to remind me that my basic needs are covered. As Erich Fromm points out in *To Have or To Be*, the most important mode isn't the "having mode,"[1] which is imbued in selfishness and narcissism, but the "being mode,"[2] which enables a person to explore his true and authentic self. So, gratitude and, more generally, mediation, help me to be calmer and more in touch with myself. Through these two important prerequisites, I feel better equipped to connect with other human beings and the natural world, both of which I care for deeply.

At the beginning of my trip, in a village in the Mexican jungle, I had read that a star would fall from the sky if a tree is cut down. Perhaps Chan Kin "Viejo" was right. We have spent so much time and energy creating artificial light that we cannot see the stars; many nocturnal insects have lost their way because of this as well. What is coming is impossible to guess, but how we approach it is entirely up to us. I became melancholic after the trip across Latin America. I felt that the bad had outweighed the good. Objectively speaking, this is true: not only did I see more destruction and death than light, but the rate of deforestation versus reforestation, the widening of the inequality gap, and the temperature rise globally were quantitative indicators that we are in trouble.

But, as I wrote up my notes each evening during the final days of my journey, I recalled how grateful I was for being able to travel freely to most places (notwithstanding being detained in Zambia and my unwillingness to enter India) and having great people in my life, together with the opportunity to see the world and be inspired by wonderful people and learn more about nature's beauty and diversity.

Most writers I know place a big wall between their personal and private lives. For me, this was impossible. Writing this precious book is an essential and authentic part of who I am, as, indeed, my previous works have been. It's as if each book is a new organ and becomes part of me. I have become friends with several of the people I met through composing both this and my first book. This is possibly a flaw, because I feel their pain a bit too much. Unbeknown to them, they are the ones that have pulled me through, because it is their tremendous courage, heart, and valor that I had been seeking for years, and the reason why I gave up my seemingly comfortable life working at a bank in 2011. I knew then that it was an illusion.

Much like the crescendo of bees who tend their queen when she needs help, or collectively fan because the hive needs to cool down in hot weather, I hope that more people will be encouraged by the wonderful people I met and act upon what they feel will be in the best interests of humanity and the natural world. I hope that as time runs out, the pace of the music of our collective efforts will increase, from lento, to allegretto, to presto, and that individual contributions will increase the colors, tones, and textures to produce an exquisite symphony of love, respect, and compassion for each other and the environment on which we all depend.

Rather than submit, acquiesce, or capitulate to the hopelessness the climate crisis warrants, I feel more motivated than ever to continue moving forward, even when many of the hard facts indicate that the trajectory has worsened since I first embarked on this journey six years ago. My commitment is not only to the people I interviewed and my manuscript, but to the natural world, which is our garden, farm, library, museum, and fountain of life.

I have faith in nature and the goodness of the human spirit. Although I distanced myself from my Catholic religion and the dogma surrounding it, I share the widely accepted belief that God is everywhere. She/he/it/they lives in every living thing. I see God when I stare through my macro lens and see a bee's eyes looking into mine. We each know that the other is there. I hope that as many species as possible will make it through the coming decades, and I also wish good tidings to all species, including each and every member of our own.

Notes

INTRODUCTION

1. Thomas Piketty, *Capital in the Twenty-First Century*. Translated by Arthur Goldhammer. (Cambridge, MA; London, England: Belknap Press of Harvard University Press, 2017).

2. United Nations, Department of Economic and Social Affairs, Population Division, *World Population Prospects 2019* (United Nations, 2019), online ed., rev. 1, https://population.un.org/wpp/Download/Files/1_Indicators%20(Standard)/EXCEL_FILES/1_Population/WPP2019_POP_F01_1_TOTAL_POPULATION_BOTH_SEXES.xlsx.

3. Paul R. Ehrlich, *The Population Bomb* (San Francisco: Sierra Club/Ballantine Books, 1968).

4. United Nations, *World Population Prospects 2019*.

5. United Nations, *World Population Prospects 2019*.

6. United Nations, *World Population Prospects 2019*.

7. David Lin et al., "Ecological Footprint Accounting for Countries: Updates and Results of the National Footprint Accounts, 2012–2018," *Resources* 7, no. 58 (2018): 1, https://www.mdpi.com/2079-9276/7/3/58.

8. United Nations, Department of Economic and Social Affairs, Population Division, *World Population Prospects: The 2012 Revision, Volume I: Comprehensive Tables ST/ESA/SER.A/336* (2013), https://population.un.org/wpp/Publications/Files/WPP2012_Volume-I_Comprehensive-Tables.pdf.

9. United Nations, Department of Economic and Social Affairs, Population Division, *World Population Prospects: The 2017 Revision, Key Findings and Advance Tables. Working Paper No. ESA/P/WP/248,* 2017, https://population.un.org/wpp/Publications/Files/WPP2017_KeyFindings.pdf.

10. Paul J. Crutzen and Eugene F. Stoermer, "The 'Anthropocene,'" *Global Change Newsletter*, 41 (2000): 17–18.

11. Judith Lewis Mernit, "How Eating Seaweed Can Help Cows to Belch Less Methane," *Yale Environment 360*, July 2, 2018, https://e360.yale.edu /features/how-eating-seaweed-can-help-cows-to-belch-less-methane.

CHAPTER 1: FROM THE JUNGLES OF MEXICO TO THE SKYSCRAPERS OF PANAMA

1. United Nations, Department of Economic and Social Affairs, Population Division, *World Population Prospects 2019* (United Nations, 2019), online ed., rev. 1, https://population.un.org/wpp/Download/Files/1_Indicators%20 (Standard)/EXCEL_FILES/1_Population/WPP2019_POP_F01_1_TO TAL_POPULATION_BOTH_SEXES.xlsx.

2. United Nations, *World Population Prospects 2019*.

3. United Nations, *World Population Prospects 2019*.

4. Thomas Malthus, *An Essay on the Principle of Population* (Digireads.com, 2013), 10, Kindle.

5. Paul R. Ehrlich and Anne H. Ehrlich, *The Population Explosion* (New York: Simon & Schuster, 1990), 17.

6. Elizabeth Kolbert, *The Sixth Extinction: An Unnatural History* (London: Bloomsbury, 2014), Kindle.

7. Hillel Steinmetz, "Noam Chomsky Warns of Sixth Extinction," *Chicago Maroon*, September 30, 2016, https://www.chicagomaroon.com/article /2016/9/30/noam-chomsky-warns-sixth-extinction/.

8. Bill Bostock, "These 12 Charts Show How the World's Population Has Exploded in the Last 200 Years," World Economic Forum, July 15, 2019, https://www.weforum.org/agenda/2019/07/populations-around-world -changed-over-the-years.

9. DinoBuzz, "What Killed the Dinosaurs?," UCMP Berkeley, accessed June 20, 2020, https://ucmp.berkeley.edu/diapsids/extinctheory.html.

10. DinoBuzz, "What Killed the Dinosaurs?"

11. Amnesty International, "The World's Refugees in Numbers," accessed June 20, 2020, https://www.amnesty.org/en/what-we-do/refugees-asylum -seekers-and-migrants/global-refugee-crisis-statistics-and-facts/.

12. Christina Nunez, "Desertification, Explained," *National Geographic*, May 31, 2019, https://www.nationalgeographic.com/environment/habitats /desertification/.

13. J. Sarukhán et al., *Capital natural de México. Síntesis: conocimiento actual, evaluación y perspectivas de sustentabilidad.* México: Comisión Nacional para el Conocimiento y Uso de la Biodiversidad, 2009, 22, http://bioteca.biodiver sidad.gob.mx/janium/Documentos/14039.pdf.

14. Alberto Martinez, in discussion with the author, Chiapas, Mexico, May 2016.

15. Martinez, in discussion with the author.

16. Martinez, in discussion with the author.

17. The Goldman Environmental Prize, "Isidro Baldenegro, 2005 Goldman Prize Recipient, North America," accessed June 22, 2020, https://www.goldmanprize.org/recipient/isidro-baldenegro/.

18. La Asociación de Cooperación al Desarrollo Integral de Huehuetenango/ Association for the Cooperation and Integral Development of Huehuetenango.

19. Braulio Palacios, "Guatemala mejora posición en 'Top 10' de países exportadores de café," Republica, December 5, 2019, https://republica.gt/2019 /12/05/guatemala-posicion-exportadores-cafe/.

20. Felix Camposeco, in discussion with the author, Huehuetenango Department, Guatemala, May 2016.

21. Consumers will pay a greater country premium for high-quality coffees from countries such as Colombia and the Democratic Republic of Congo (DRC).

22. Urías Gamarro, "Guatemaltecos enviaron remesas por más de US$1 millón por hora en 2019," *Prensa Libre*, January 9, 2020, https://www.prensalibre .com/economia/guatemaltecos-enviaron-remesas-por-mas-de-us1-millon -por-hora-en-2019/.

23. Palacios, "Guatemala mejora posición en 'Top 10' de países exportadores de café."

24. Carlos, in discussion with the author, Huehuetenango Department, Guatemala, May 2016.

25. United Nations, Convention on Climate Change, *The Kyoto Protocol —Status of Ratification*, accessed June 20, 2020, https://unfccc.int/process/the -kyoto-protocol/status-of-ratification.

26. United Nations, *Kyoto Protocol to the United Nations Framework Convention on Climate Change* (New York: United Nations, 1998), 19, https://unfccc .int/resource/docs/convkp/kpeng.pdf.

27. Oxfam International, *An Economy for the 1%: How Privilege and Power in the Economy Drive Extreme Inequality and How This Can Be Stopped* (Oxford: Oxfam GB), 1, https://s3.amazonaws.com/oxfam-us/www/static/media/files /bp210-economy-one-percent-tax-havens-180116-en_0.pdf.

28. Oxfam International, *An Economy for the 1%*, 2.

29. Luke Harding, "Panama Papers Investigation Wins Pulitzer Prize," *The Guardian*, April 11, 2017, https://www.theguardian.com/world/2017/apr/11 /panama-papers-investigation-wins-pulitzer-prize.

30. Süddeutsche Zeitung, "We Don't Regret Anything," accessed June 20, 2020, https://panamapapers.sueddeutsche.de/articles/57162ccca1bb8d3c3495 bc5e/.

31. BBC News, "Panama Papers: What Happened Next?," *BBC*, December 26, 2016, https://www.bbc.com/news/world-38319026.

32. Paul Karp, "Coalition Says Panama Papers Sparked Action against More Than 100 Australian Taxpayers," *The Guardian*, September 6, 2016, https://www.theguardian.com/australia-news/2016/sep/06/coalition-says -panama-papers-sparked-action-against-more-than-100-australian-taxpayers.

33. Juliette Garside, "Panama Papers: European Parliament Opens Inquiry," *The Guardian*, September 27, 2016, https://www.theguardian.com/world /2016/sep/27/panama-papers-inquiry-opens-at-european-parliament.

34. Deutsche Welle (DW), "Tax Authorities in Denmark Buy 'Panama Papers' Evidence," *DW*, September 29, 2016, https://www.dw.com/en/tax -authorities-in-denmark-buy-panama-papers-evidence/a-35928320.

35. Foo Yun Chee and Kirstin Ridley, "EU Fines Barclays, Citi, JP Morgan, MUFG and RBS $1.2 Billion for FX Rigging," *Reuters*, May 16, 2019, https:// www.reuters.com/article/us-eu-antitrust-banks/eu-fines-barclays-citi-jp -morgan-mufg-and-rbs-12-billion-for-fx-rigging-idUSKCN1SM0XS.

36. James McBride, "Understanding the Libor Scandal," *Council on Foreign Relations*, October 12, 2016, https://www.cfr.org/backgrounder/understanding -libor-scandal.

37. David Teather, "Enron Accountant Pleads Guilty," *The Guardian*, December 29, 2005, https://www.theguardian.com/business/2005/dec/29/cor poratefraud.enron.

38. Ashley Cowburn, "David Cameron Offshore Fund: Video of Four Times PM Condemned 'Immoral' Tax Avoidance," *The Independent*, April 8, 2016, https://www.independent.co.uk/news/uk/politics/david-cameron -panama-papers-tax-avoidance-father-offshore-investment-fund-ian-cam eron-blairmore-a6974546.html.

39. Maria Gallucci, "At Last, the Shipping Industry Begins Cleaning up Its Dirty Fuels," *Yale Environment 360*, June 28, 2018, https://e360.yale.edu /features/at-last-the-shipping-industry-begins-cleaning-up-its-dirty-fuels.

40. Ed Struzik, "Shipping Plans Grow as Arctic Ice Fades," *Yale Environment 360*, November 17, 2016, https://e360.yale.edu/features/cargo_shipping _in_the_arctic_declining_sea_ice.

CHAPTER 2: COLOMBIAN ENERGY

1. Mauricio Toro, in discussion with the author, Guasca, Colombia, 2016.

2. Toro, in discussion with the author.

3. United Nations, Department and Economic and Social Affairs, Population Division. *The World's Cities in 2018—Data Booklet* (New York: United

Nations, 2018), 4 and 26, https://www.un.org/en/events/citiesday/assets/pdf /the_worlds_cities_in_2018_data_booklet.pdf.

4. Andres Moreno, in discussion with the author, multiple locations in Colombia, June 2016.

5. Moreno, in discussion with the author.

6. Moreno, in discussion with the author.

7. IEA, *Solar Energy Perspectives* (Paris: IEA, 2011), https://www.iea.org /reports/solar-energy-perspectives.

8. BBC News, "Solar Impulse Completes Historic Round-The-World Trip," *BBC*, July 26, 2016, https://www.bbc.com/news/science-environment -36890563.

9. Pensamiento y Acción Social, and Arbeitsgruppe Schweiz Kolumbien, *Shadow Report on the Sustainability of Glencore's Operations in Colombia* (Bogota, 2015), 17–18.

10. Dennis Kennedy, "Environmental Degradation and Disrupted Social Fabric in the Tar Creek Basin" (PhD diss., Oklahoma State University, 2008), 119.

11. Bruce Douglas, "Anger Rises as Brazilian Mine Disaster Threatens River and Sea with Toxic Mud," *The Guardian*, November 22, 2015, https:// www.theguardian.com/business/2015/nov/22/anger-rises-as-brazilian-mine -disaster-threatens-river-and-sea-with-toxic-mud.

12. Transparency International, *Corruption Perceptions Index 2019*, accessed June 20, 2020, 3, https://images.transparencycdn.org/images/2019_CPI_Report _EN_200331_141425.pdf.

13. Reporters Without Borders (RSF), "2020 World Press Freedom Index," accessed June 20, 2020, https://rsf.org/en/ranking.

14. Global Witness, "On Dangerous Ground," June 20, 2016, https://www .globalwitness.org/en/campaigns/environmental-activists/dangerous-ground/.

15. Reuters, "Colombia Coal Production Up 6.5% in First Quarter," May 22, 2019, https://www.reuters.com/article/us-colombia-coal/colombia-coal-pro duction-up-6-5-in-first-quarter-idUSKCN1SS2OA.

16. Heffa Schücking, *Banking on Coal* (Urgewald, BankTrack, CEE Bank-watch Network and Polska Zielona Sieć, 2013), https://www.banktrack.org /download/banking_on_coal/banking_on_coal_updated.pdf.

17. Pseudonym used to protect anonymity.

18. Jaime, in discussion with the author, El Hatillo, Colombia, June 2016.

19. Jaime, in discussion with the author.

20. Jaime, in discussion with the author.

21. Tatiana Rojas, in discussion with the author, El Hatillo, Colombia, June 2016.

22. Vale, *A Year of Extraordinary Performance: Annual Report 2010* (Vale S. A., 2011), 59, http://www.vale.com/EN/investors/information-market/annual-reports/20f/20FDocs/20F_2010_i.pdf.
23. Jaime, in discussion with the author.

CHAPTER 3: TWO GREAT FORESTS

1. Although cocaine is produced in smaller quantities than beef, which will be discussed in chapter 5, both "products" contribute to an increased carbon footprint due to deforestation. An additional environmental complication of cocaine is the chemical process used to convert coca leaves into paste.
2. Charles Parkinson, "The Flow of Drugs and Blood in the Amazon Tri-Border Region," *InSight Crime*, April 4, 2014, https://www.insightcrime.org/news/analysis/the-flow-of-drugs-and-blood-in-the-amazon-tri-border-region/.
3. David Gagne, "How Drug Traffickers Operate in Peru's Amazon Region," *InSight Crime*, April 7, 2015, https://www.insightcrime.org/news/brief/how-drug-traffickers-operate-peru-amazon/.
4. Ministerio da Saude, "Risk Assessment—Olympic and Paralympic Games," Brazil's Ministry of Health, accessed June 23, 2020, http://www.saude.gov.br/images/pdf/2016/agosto/05/MA---2.pdf.
5. Ana Mano, "Brazil's Beef Export Hit Record, Prospects Bright on China Demand," *Reuters*, December 10, 2019, https://www.reuters.com/article/us-brazil-beef/brazils-beef-export-hit-record-prospects-bright-on-china-demand-idUSKBN1YE1TS.
6. Gidon Eshel et al., "Land, Irrigation Water, Greenhouse Gas, and Reactive Nitrogen Burdens of Meat, Eggs, and Dairy Production in the United States," *Proceedings of the National Academy of Sciences of the United States of America* 111, no. 33 (Summer 2014), https://doi.org/10.1073/pnas.1402183111.
7. WWF, "Amazon Deforestation," accessed June 22, 2020, https://wwf.panda.org/our_work/forests/deforestation_fronts2/deforestation_in_the_amazon/.
8. Sam Cowie, "The Jungle Metropolis: How Sprawling Manaus Is Eating into the Amazon," *The Guardian*, July 23, 2019, https://www.theguardian.com/cities/2019/jul/23/the-jungle-metropolis-how-sprawling-manaus-is-eating-into-the-amazon.
9. Dias de Oliveira et al., "Ethanol as Fuel: Energy, Carbon Dioxide Balances, and Ecological Footprint," *BioScience* 55, no. 7 (July 2005): 601, https://doi.org/10.1641/0006-3568(2005)055[0593:EAFECD]2.0.CO;2.

10. Nicolle Monteiro de Castro, "Brazil Center-South Cumulative Ethanol Production in Crop Year 2019–20 Rises 6.60% Year on Year: UNICA," *S&P Global*, January 27, 2020, https://www.spglobal.com/platts/en/market-insights/latest-news/agriculture/012720-brazil-center-south-cumulative-ethanol-production-in-crop-year-2019-20-rises-660-year-on-year-unica.

11. The Nature Conservancy, "The Atlantic Forest," accessed June 22, 2020, https://www.nature.org/en-us/get-involved/how-to-help/places-we-protect/atlantic-forest/.

12. Nature, "Local Spotlight. Rio de Janeiro, Brazil—Measuring Biodiversity and Ecological Integrity Benefits," accessed June 22, 2020, https://www.nature.org/content/dam/tnc/nature/en/photos/LocalSpotlight_RioDe Janeiro_Brazil.pdf.

13. WWF, "Second Only to the Amazon," accessed June 22, 2020, https://wwf.panda.org/knowledge_hub/where_we_work/atlantic_forests/.

14. IMDb, "The Salt of the Earth," (2014), accessed June 22, 2020, https://www.imdb.com/title/tt3674140/?ref_=ttmc_mc_tt.

CHAPTER 4: INGENIOUS SHANTYTOWNS

1. Fabio Miranda, in discussion with the author, Jardim Ângela, Brazil, July 2016.

2. Miranda, in discussion with the author.

3. Miranda, in discussion with the author.

4. Miranda, in discussion with the author.

5. Miranda, in discussion with the author.

6. Miranda, in discussion with the author.

7. World Health Organization, "Brazil," accessed June 25, 2020, https://www.who.int/diabetes/country-profiles/bra_en.pdf.

8. Miranda, in discussion with the author.

9. Miranda, in discussion with the author.

10. Marcio Weber, in discussion with the author, Asunción, Paraguay, July 2016.

11. Weber, in discussion with the author.

12. Favio Chávez, in discussion with the author, Asunción, Paraguay, July 2016.

13. Chávez, in discussion with the author.

14. Chávez, in discussion with the author.

15. Chávez, in discussion with the author.

16. Chávez, in discussion with the author.

CHAPTER 5: THE OPEN VEINS OF LATIN AMERICA

1. Pseudonym used to protect anonymity.
2. Pseudonym used to protect anonymity.
3. Pseudonym used to protect anonymity.
4. Chloe, in discussion with the author, Bajo Hondo, Argentina, July 2016.
5. Chloe, in discussion with the author.
6. Chloe, in discussion with the author.
7. Chloe, in discussion with the author.
8. Chloe, in discussion with the author.
9. David Cox, *Dirty Secrets, Dirty War: Buenos Aires, Argentina, 1976–1983: The Exile of Robert J. Cox* (Charleston: Evening Post, 2008), 20.
10. Chloe, in discussion with the author.
11. Chloe, in discussion with the author.
12. ANRed, "Otro ataque de Manaos a las comunidades campesinas en Santiago del Estero," July 5, 2016, https://www.anred.org/2016/07/05/otro -ataque-de-manaos-a-las-comunidades-campesinas-en-santiago-del-estero/; La Izquierda Diario, "Santiago del Estero: brutal ataque contra la Comunidad Bajo Fondo por parte de la empresa Manaos," September 24, 2016, http:// www.laizquierdadiario.com/Santiago-del-Estero-brutal-ataque-contra-la -Comunidad-Bajo-Fondo-por-parte-de-la-empresa-Manaos.
13. Walter Pengue, "The Impact of Soybean Expansion in Argentina," *Grain*, September 24, 2001, https://www.grain.org/es/article/entries/292-the -impact-of-soybean-expansion-in-argentina.
14. Ayn Rand, *The Fountainhead* (New York: New American Library, 2016), Kindle, loc. 14707.
15. Eduardo Galeano, *Open Veins of Latin America: Five Centuries of the Pillage of a Continent*, translated by Cedric Belfrage (New York: Monthly Review Press, 1997), Kindle, 1.
16. BBC News, "Brazil Mensalao Trial: Former Chief of Staff Jailed," BBC, November 16, 2013, https://www.bbc.com/news/world-latin-america -24967116.
17. Jason Beaubien, "Tijuana Violence Likely to Continue in 2009," *NPR*, February 1, 2009, https://www.npr.org/templates/story/story.php?storyId =100063216.
18. Ioan Grillo, "Inside El Salvador's 'War Without Sense,'" *Time*, July 23, 2015, https://time.com/3966900/el-salvador-gangs-violence/.
19. Anggy Polanco, "Venezuela Probes Reports of Prison Cannibal Deaths," *Reuters*, October 17, 2016, https://www.reuters.com/article/us-venezuela-can nibalism-idUSKBN12H1SZ.

20. Justin Rohrlich, "This Is How Much It Costs to Be Smuggled over the US Border," *Quartz*, June 10, 2019, https://qz.com/1632508/this-is-how-much-it-costs-to-cross-the-us-mexico-border-illegally/.

21. United Nations, Department of Economic and Social Affairs, Population Division. *World Population Prospects 2019* (United Nations, 2019), online ed., rev. 1, https://population.un.org/wpp/Download/Files/1_Indicators%20(Standard)/EXCEL_FILES/1_Population/WPP2019_POP_F01_1_TOTAL_POPULATION_BOTH_SEXES.xlsx.

22. The Conversation, "A Decade of Murder and Grief: Mexico's Drug War Turns Ten," December 11, 2016, https://theconversation.com/a-decade-of-murder-and-grief-mexicos-drug-war-turns-ten-70036.

23. Alberto Morales, "Calderon and Pena Nieto Spent $675 Million on Planes and Helicopters," *El Universal*, November 24, 2019, https://www.eluniversal.com.mx/english/calderon-and-pena-nieto-spent-675-million-planes-and-helicopters.

24. This was calculated by subtracting the "Caribbean region" value from the "Latin America and Caribbean" value to arrive at the value for "Latin America" in the "estimates" variant of the UN's *World Population Prospects 2019* data set.

25. United Nations, *World Population Prospects 2019*.

26. United Nations, *World Population Prospects 2019*.

27. IIRSA, "Taller Sobre Agendas Cartográficas," Buenos Aires, Argentina, November 11, 2011, http://www.iirsa.org/admin_iirsa_web/Uploads/Documents/geo11_baires11_agenda_cartografica.pdf.

28. The rankings for South American countries from a list of 180 countries was as follows, in descending order (least corrupt first): 21 Uruguay, 26 Chile, 66 Argentina, 70 Suriname, 85 Guyana, 93 Ecuador, 96 Colombia, 101 Peru, 106 Brazil, 123 Bolivia, 173 Venezuela. The data for French Guiana were unavailable. As a point of reference, Canada and the UK both ranked 12, with two others, and the United States 23. Source: Transparency International. *Corruption Perceptions Index 2019* (Transparency International, 2019), accessed June 20, 2020. https://images.transparencycdn.org/images/2019_CPI_Report_EN_200331_141425.pdf.

29. Galeano, *Open Veins of Latin America*, 223.

30. Galeano, *Open Veins of Latin America*, 223.

31. Santander, "Support for Santander Mexico Customers on Account of COVID-19," March 26, 2020, https://www.santander.com/content/dam/santander-com/en/documentos/actualidad/2020/03/no-2020-03-26-support-for-santander-mexico-customers-on-account-of-covid-19-en.pdf.

CHAPTER 6: AMERICAN POWER

1. Jordan Golson, "The New Tesla Model S P100D Is Faster Than All These Cars," *The Verge*, August 23, 2016, https://www.theverge.com/2016/8/23/12615304/tesla-model-s-p100d-fastest-acceleration.

2. Terrence Collingsworth, "Class Complaint for Injunctive Relief and Damages," *International Rights Advocates*, December 15, 2019, http://iradvocates.org/sites/iradvocates.org/files/stamped%20-Complaint.pdf.

3. John M. Deutch, *The Crisis in Energy Policy* (Cambridge, MA; London, UK: Harvard University Press, 2011), 14.

4. BBC News, "Former VW Boss Charged over Diesel Emissions Scandal," *BBC*, April 15, 2019, https://www.bbc.com/news/47937141.

5. BBC News, "Mitsubishi Motors Admits Using Wrong Tests since 1991," *BBC*, April 26, 2016, https://www.bbc.com/news/business-36137719.

6. Daniel Cusick, "Miami Is the "Most Vulnerable" Coastal City Worldwide," *Scientific American*, February 4, 2020, https://www.scientificamerican.com/article/miami-is-the-most-vulnerable-coastal-city-worldwide/.

7. BBC News, "Renewable Energy Capacity Overtakes Coal," *BBC*, October 25, 2016, https://www.bbc.com/news/business-37767250.

8. Maria Ceron, in discussion with the author, Miami, November 2016.

9. Ceron, in discussion with the author.

10. Ceron, in discussion with the author.

11. Parts per million (ppm) is an ambiguous scale which gives an indication of how concentrated a substance is. In this case, CO_2 had never passed this level in millions of years (source: Nicola Jones, "How the World Passed a Carbon Threshold and Why It Matters," *Yale Environment 360*, January 26, 2017, https://e360.yale.edu/features/how-the-world-passed-a-carbon-threshold-400ppm-and-why-it-matters). Moreover, the ppm level gives an indication of a warming planet. The higher the CO_2 level, the greater its warming potential and, therefore, the warmer the planet will become. The more that fossil fuels are burned, trees felled, and other carbon sources released into the atmosphere, the greater this number becomes.

12. Jones, "How the World Passed a Carbon Threshold and Why It Matters."

13. Ceron, in discussion with the author.

14. Ceron, in discussion with the author.

15. Megan Greenwalt, "A Look at the Largest Landfill Gas-To-Energy Project in Georgia," July 27, 2016, https://www.waste360.com/gas-energy/look-largest-landfill-gas-energy-project-georgia.

16. Aaron Scranton, written exchanges, 2016.

17. Scranton, written exchanges.

18. Scranton, written exchanges.

19. Scranton, written exchanges.

20. Gardiner Harris, "Rex Tillerson Is Confirmed as Secretary of State amid Record Opposition," *New York Times*, February 1, 2017, https://www.nytimes .com/2017/02/01/us/politics/rex-tillerson-secretary-of-state-confirmed.html.

21. Steve Nelson, in discussion with the author, New York City, November 2016.

22. Joseph Lipari, in discussion with the author, New York City, November 2016.

23. Nyc.gov, "Climate Week: Solar Power in NYC Nearly Quadrupled since Mayor de Blasio Took Office and Administration Expands Target," *NYC*, September 23, 2016, https://www1.nyc.gov/office-of-the-mayor/news/767-16 /climate-week-solar-power-nyc-nearly-quadrupled-since-mayor-de-blasio -took-office-and.

24. Ernest Hemingway, *To Have and Have Not* (New York: Scribner, 1937).

25. Oxfam, "Poverty in the USA," accessed June 23, 2020, https://policy -practice.oxfamamerica.org/work/poverty-in-the-us/.

26. BBC News, "Blogger Travels from Sheffield to Essex via Berlin to Save Cash," *BBC*, January 28, 2016, https://www.bbc.com/news/uk-england -35424393.

27. BBC News, "Sweden Sees Rare Fall in Air Passengers, as Flight-Shaming Takes Off," *BBC*, January 10, 2020, https://www.bbc.com/news/world -europe-51067440.

CHAPTER 7: FOOD IN THE CITY

1. History, "The Pilgrims," November 21, 2019, https://www.history.com /topics/colonial-america/pilgrims.

2. Hyper-local refers to an extremely short distance between farm and point of sale, compared with the hundreds or thousands of miles that food may have to travel (chapter 15).

3. Julie McMahon, in discussion with the author, New York City, November 2016.

4. McMahon, in discussion with the author.

5. Orangutan Foundation, "The Effects of Palm Oil," accessed June 22, 2020, https://orangutan.org/rainforest/the-effects-of-palm-oil/.

6. WWF, "8 Things to Know about Palm Oil," January 17, 2020, https:// www.wwf.org.uk/updates/8-things-know-about-palm-oil.

7. Zoe Wood, "Fast Food Chains Prosper as Cash-Strapped Consumers Shun Retailers," *The Guardian*, January 29, 2012, https://www.theguardian .com/business/2012/jan/29/fast-food-chains-prosper-retailers-under-pressure.

8. National Institute of Diabetes and Digestive and Kidney Diseases (NIDDK), "Overweight & Obesity Statistics," August 2017, https://www.niddk.nih.gov/health-information/health-statistics/overweight-obesity?dkrd=hispt0880.

9. The Michigan Urban Farming Initiative, "What's Going on at MUFI?," accessed June 23, 2020, https://www.miufi.org/projects.

10. Feedback, "Feeding the 5,000," accessed June 23, 2020, https://feedbackglobal.org/campaigns/feeding-the-5000/.

11. Susan Daugherty, "Tristram Stuart: Waging War against Global Food Waste," *National Geographic*, October 16, 2014, https://www.nationalgeographic.com/news/2014/10/141016-food-waste-tristram-stuart-emerging-explorer-hunger-charity-ngfood/.

12. Tristram Stuart, video call, November 2016.

13. Stuart, video call.

14. Stuart, video call.

15. Vanesa de Blas, in discussion with the author, London, UK, December 2016.

16. de Blas, in discussion with the author.

17. Trussell Trust, "About," accessed June 22, 2020, https://www.trusselltrust.org/about/.

18. Trussell Trust, "Food Bank Statistics for Previous Financial Years with Regional Breakdown 2018–2019," accessed June 20, 2020, https://www.trusselltrust.org/news-and-blog/latest-stats/end-year-stats/#fy-2018-2019.

19. Trussell Trust, "Food Bank Statistics for Previous Financial Years with Regional Breakdown 2017–2018," accessed June 22, 2020, https://www.trusselltrust.org/news-and-blog/latest-stats/end-year-stats/#fy-2017-2018.

20. Eurofoodbank, "The Global Context of Food Waste," accessed June 23, 2020, https://www.eurofoodbank.org/en/food-waste.

21. Matt Payton, "French Supermarkets Banned from Throwing away and Spoiling Unsold Food," *The Independent*, February 5, 2016, https://www.independent.co.uk/news/world/europe/french-law-bans-supermarkets-throwing-away-and-spoiling-unsold-food-giving-them-to-food-banks-and-a6855371.html.

22. Stephanie Kirchgaessner, "Italy Tackles Food Waste with Law Encouraging Firms to Donate Food," *The Guardian*, August 3, 2016, https://www.theguardian.com/world/2016/aug/03/italy-food-waste-law-donate-food.

23. Jamie Merrill, "Ugly Fruit Gets a Makeover as Store Cuts Waste," *The Independent*, May 31, 2014, https://www.independent.co.uk/life-style/food-and-drink/news/ugly-fruit-gets-makeover-store-cuts-waste-9466943.html.

24. Michelin Guide, "Michelin Guide Denmark 2020: Relae," accessed June 22, 2020, https://guide.michelin.com/en/capital-region/kobenhavn/restaurant/relae.

25. The World's 50 Best Restaurants, "The World's 50 Best Restaurants 2017," 2017, https://www.theworlds50best.com/previous-list/2017.

26. The World's 50 Best Restaurants, "Sustainable Restaurant Award," 2019, https://www.theworlds50best.com/awards/sustainable-restaurant-award.

27. Lisa Lov, in discussion with the author, Copenhagen, Denmark, December 2016.

28. Lov, in discussion with the author.

29. Lov, in discussion with the author.

30. Lars Jacobsen, written exchanges, 2016 and 2017.

31. Jacobsen, written exchanges.

32. Zlata Rodionova, "Denmark Reduces Food Waste by 25% in Five Years with the Help of One Woman—Selina Juul," *The Independent*, February 28, 2017, https://www.independent.co.uk/news/business/news/denmark-reduce-food-waste-25-per-cent-five-years-help-selina-juul-scandanavia-a7604061.html.

33. Stop Wasting Food, "Stop Wasting Food," accessed June 23, 2020, https://stopwastingfoodmovement.org/.

CHAPTER 8: CITIES OF THE FUTURE

1. Middelgrunden, "About Middelgrunden Wind Cooperative," accessed June 25, 2020, http://www.middelgrunden.dk/middelgrunden/?q=en/node/35.

2. Paul A. Lynn, *Onshore and Offshore Wind Energy: An Introduction* (Chichester: Wiley, 2012), 167.

3. Lynn, *Onshore and Offshore Wind Energy*, 167.

4. Ramboll, "Anholt Offshore Wind Farm—Denmark's Largest Offshore Wind Farm," accessed June 25, 2020, https://ramboll.com/projects/re/anholt-offshore-wind-farm.

5. World Wind Energy Association (WWEA), "Denmark," *WWEA Policy Paper Series* (April 2018): 1, http://www.wwindea.org/wp-content/uploads/2018/06/Denmark_full.pdf.

6. Martin Hansen, in discussion with the author, Copenhagen, Denmark, December 2016.

7. Hansen, written exchanges, 2016.

8. Arthur Neslen, "Wind Power Generates 140% of Denmark's Electricity Demand," *The Guardian*, July 10, 2015, https://www.theguardian.com/environment/2015/jul/10/denmark-wind-windfarm-power-exceed-electricity-demand.

9. Neslen, "Wind Power Generates 140% of Denmark's Electricity Demand."

10. Hansen, written exchanges, 2016.

11. Laurie Winkless, "Dutch Trains Are Now Powered by Wind," *Forbes*, January 12, 2017, https://www.forbes.com/sites/lauriewinkless/2017/01/12/dutch-trains-are-now-powered-by-wind/#631148752d29.

12. Katie Forster, "Paris Bans Cars with Even Plates in Third Day of Transport Restrictions to Ease Air Pollution," *The Independent*, December 8, 2016, https://www.independent.co.uk/news/world/europe/paris-banned-cars-even-number-plates-pollution-latest-public-transport-air-smog-bans-a7462621.html.

13. Given the notoriously polluted air in Mexico City, the local government has invoked a state of environmental contingency on numerous occasions when levels passed certain limits. In order to alleviate the pollution, automobiles with certain number plates were not allowed to be used, and the metro system was free for all passengers. The government also has a cell phone application that provides information about air quality, giving recommendations about the suitability of outdoor activities.

14. City of Amsterdam, "Policy: Clean Air," accessed June 25, 2020, https://www.amsterdam.nl/en/policy/sustainability/clean-air/.

15. Andrea Lo, "The Netherlands Is Paying People to Cycle," *CNN*, December 21, 2018, https://edition.cnn.com/travel/article/netherlands-cycling/index.html.

16. ITSInternational, "Barcelona's Bike Share Scheme a Lifesaver," January 26, 2012, https://www.itsinternational.com/feature/barcelonas-bike-share-scheme-life-saver.

17. Mobike, "What Is Mobike," accessed June 25, 2020, https://mobike.com/global/faq.

18. Euronews, "'Ryanair Is the New Coal' as It Becomes First Airline in EU's Top Ten Biggest Polluters," April 3, 2019, https://www.euronews.com/2019/04/02/ryanair-is-the-new-coal-airline-is-eu-s-10th-biggest-polluter.

19. Alejandra Borunda, "Most Dire Projection of Sea-Level Rise Is a Little Less Likely, Reports Say," *National Geographic*, February 6, 2019, https://www.nationalgeographic.com/environment/2019/02/antarctic-greenland-ice-melt-less-bad/.

20. Shoshana Dubnow, "Report Indicates Increase of Cocaine Usage in EU, Links to Social Media," *Euronews*, June 6, 2019, https://www.euronews.com/2019/06/06/report-indicates-increase-of-cocaine-usage-in-eu-links-to-social-media.

21. European Monitoring Centre for Drugs and Drug Addiction (EMCDDA), "Drug Use Prevalence and Trends," accessed June 25, 2020, https://www.emcdda.europa.eu/publications/edr/trends-developments/2017/html/prevalence-trends/cocaine_en.

22. United Nations Office on Drugs and Crime (UNODC), "Heroin and Cocaine Prices in Europe and USA," *United Nations*, accessed June 24, 2020, https://dataunodc.un.org/drugs/heroin_and_cocaine_prices_in_eu_and_usa -2017.

CHAPTER 9: CROSSING EUROPE

1. May Bulman, "Calais Jungle Closed: Hundreds of Children Remain Unaccounted for Despite Official Closure," *The Independent*, October 26, 2016, https://www.independent.co.uk/news/world/europe/calais-jungle-closed-chil dren-unaccounted-official-closure-camp-refugees-a7381626.html.
2. Roz McGregor, in discussion with the author, London, United Kingdom, December 2016.
3. McGregor, in discussion with the author.
4. McGregor, in discussion with the author.
5. McGregor, in discussion with the author.
6. UNHCR, "79.5 Million Forcibly Displaced People Worldwide at the End of 2019," June 18, 2020, https://www.unhcr.org/figures-at-a-glance.html.
7. UNHCR, "79.5 Million Forcibly Displaced People Worldwide at the End of 2019."
8. Olympic.org, "The Refugee Olympic Team, A Symbol of Hope," April 5, 2017, https://www.olympic.org/news/the-refugee-olympic-team-a-symbol -of-hope.
9. BBC Sport, "Rio Olympics 2016: Refugee Olympic Team Competed As 'Equal Human Beings,'" *BBC*, August 21, 2016, https://www.bbc.com /sport/olympics/37037273.
10. Human Rights Watch, "Off the Radar," October 18, 2014, https:// www.hrw.org/report/2014/10/18/radar/human-rights-tindouf-refugee-camps.
11. Pseudonym used to protect anonymity.
12. Saba, in discussion with the author, Bratislava, Slovakia, December 2016.
13. Saba, in discussion with the author.
14. Saba, in discussion with the author.
15. Barhum Nakhlé, in discussion with the author, Bratislava, Slovakia, December 2016.
16. Nakhlé, in discussion with the author.
17. Nakhlé, in discussion with the author.
18. Nima Elbagir et al., "The US Cleared the Way for a New Arms Sale to the UAE, Despite Evidence It Violated the Last One," *CNN*, May 22, 2020,

https://edition.cnn.com/2020/05/22/world/state-department-uae-arms-sale-yemen-intl/index.html.

19. Lizzie Dearden, "Billions of Pounds Worth of Weapons Have Been Licensed by the UK to Saudi Arabia since Start of Yemen War," *The Independent*, August 21, 2019, https://www.independent.co.uk/news/uk/home-news/saudi-arabia-arms-sales-uk-trade-yemen-war-crimes-a9073836.html.

20. BBC News, "Prince Charles Takes Part in Saudi Arabian Sword Dance," *BBC*, February 19, 2014, https://www.bbc.com/news/uk-26252091.

21. Paulina Cachero, "US Taxpayers Have Reportedly Paid an Average of $8,000 Each and Over $2 Trillion Total for the Iraq War Alone," *Business Insider*, February 6, 2020, https://www.businessinsider.com/us-taxpayers-spent-8000-each-2-trillion-iraq-war-study-2020-2?r=MX&IR=T.

22. At the time, Nigel Farage was the leader of the United Kingdom Independence Party (UKIP), a right-wing populist political party.

23. Laura Jaitman and Stephen Machin, "Crime and Immigration: New Evidence from England and Wales," *IZA J Migration* 2, no. 19 (2013), https://doi.org/10.1186/2193-9039-2-19.

24. Zygmunt Bauman, *Liquid Fear* (Cambridge: Polity Press, 2006).

25. Amnesty International, "The World's Refugees in Numbers," *Amnesty International*, accessed June 20, 2020, https://www.amnesty.org/en/what-we-do/refugees-asylum-seekers-and-migrants/global-refugee-crisis-statistics-and-facts/.

26. Rachel Hosie, "What's It Like to Go on Holiday to Turkey Right Now?" *The Independent*, August 25, 2017, https://www.independent.co.uk/travel/europe/turkey-holidays-whats-it-like-bodrum-terror-attack-political-coup-summer-tourism-a7908636.html.

27. BBC News, "Istanbul Reina Attack Suspect Says Nightclub Was Chosen at Random," *BBC*, January 19, 2017, https://www.bbc.com/news/world-europe-38673154.

28. IAEA, "Country Nuclear Power Profiles: Turkey," accessed June 25, 2020, https://www-pub.iaea.org/MTCD/Publications/PDF/cnpp2018/countryprofiles/Turkey/Turkey.htm.

29. Michael Day, "Costa Concordia Trial: Was Captain Francesco Schettino Really the Only One at Fault for the Disaster?," *The Independent*, February 15, 2015, https://www.independent.co.uk/news/world/europe/costa-concordia-trial-was-captain-francesco-schettino-really-only-one-fault-disaster-10046725.html.

30. Roswitha Oschmann, "Sea Watch Captain Speaks in Bad Honnef," *General Anzeiger*, April 5, 2019, https://www.general-anzeiger-bonn.de/region/siebengebirge/bad-honnef/sea-watch-kapitaenin-spricht-in-bad-honnef_aid-44029639.

31. Jessica Elgot, "Nigel Farage: two of my children have German passports," *The Guardian*, April 18, 2018, https://www.theguardian.com /politics/2018/apr/18/nigel-farage-admits-his-children-hold-german-as-well -as-uk-passports.

32. Gov.uk, "UK Becomes First Major Economy to Pass Net Zero Emissions Law," June 27, 2019, https://www.gov.uk/government/news/uk-becomes -first-major-economy-to-pass-net-zero-emissions-law#:~:text=The%20 UK%20today%20became%20the,80%25%20reduction%20from%201990%20 levels.

33. Marcus Wacket, "Germany to Phase out Coal by 2038 in Move away from Fossil Fuels," *Reuters*, January 25, 2019, https://www.reuters.com/article /us-germany-energy-coal/germany-to-phase-out-coal-by-2038-in-move -away-from-fossil-fuels-idUSKCN1PK04L.

34. Jan Strupczewski and Gabriela Baczynska, "EU Leaves Poland out of 2050 Climate Deal after Standoff," *Reuters*, December 11, 2019, https://www .reuters.com/article/us-climate-change-eu/eu-leaves-poland-out-of-2050 -climate-deal-after-standoff-idUSKBN1YG01I.

35. Ivana Kottasová, "Norway Says Its New Giant Oil Field is Actually Good for the Environment. Critics Call It Climate Hypocrisy," *CNN*, January 19, 2020, https://amp.cnn.com/cnn/2020/01/19/business/norway-oil-field -climate-change-intl/index.html.

36. Oslo Reuters, "Norway Wealth Fund Grows to Record 10 Trillion Crowns," *Reuters*, October 25, 2019, https://www.reuters.com/article/us-norway -swf-record/norway-wealth-fund-grows-to-record-10-trillion-crowns-idUS KBN1X41AO.

37. Columbia University, "EPA Revokes California's Authority to Set Climate-Protective Vehicle Emissions Standards," accessed June 25, 2020, https://climate.law.columbia.edu/content/epa-revokes-californias-authority -set-climate-protective-vehicle-emissions-standards.

38. Columbia University, "President Issues Executive Order Revoking Federal Sustainability Plan," accessed June 25, 2020, https://climate.law.columbia .edu/content/president-issues-executive-order-revoking-federal-sustainability -plan-0.

39. The Canadian Press, "U.S. Methane Regulation Reversal Widens Its Competitive Advantage over Canada, Energy Industry Group Says," *CBC*, August 30, 2019, https://www.cbc.ca/news/canada/calgary/methane-regulations -canada-usa-alberta-oil-gas-capp-emissions-trump-1.5265988.

40. Andrew Buncombe, "US Tells UN It Is Officially Pulling out of Paris Agreement on Climate Change," *The Independent*, November 4, 2019, https:// www.independent.co.uk/news/world/americas/us-politics/paris-agreement -climate-change-us-un-accord-pact-a9185171.html.

41. During a speech on June 16, 2015, Donald Trump said, "When Mexico sends its people, they're not sending their best. They're not sending you. They're not sending you. They're sending people that have lots of problems, and they're bringing those problems with us. They're bringing drugs. They're bringing crime. They're rapists. And some, I assume, are good people." Source: Time Staff, "Here's Donald Trump's Presidential Announcement Speech," *Time*, June 16, 2015, https://time.com/3923128/donald-trump-announcement -speech/.

42. Andrea Buring and Alice Tidey, "Germany Offers Year of Rent to Asylum Seekers Who Return Home," *Euronews*, November 27, 2018, https:// www.euronews.com/2018/11/27/germany-offers-year-of-rent-to-asylum -seekers-who-return-home.

43. Erich Fromm, *The Art of Loving* (New York: HarperCollins, 1956).

44. Erich Fromm, *To Have or To Be* (New York: Open Road Integrated Media, 2013), Kindle.

45. Elia E. Martinez Mercado, in discussion with the author, Mexico City, July 2017.

46. Martinez Mercado, in discussion with the author.

47. Martinez Mercado, in discussion with the author.

48. Paty Fuentes, in discussion with the author, Mexico City, July 2017.

CHAPTER 10: REFLECTING AND REVIVING

1. Peat moss is an effective addition to make flowers grow stronger, but to the detriment of ecosystems where the peat is sourced. Source: Jesse Vernon Trail, "The Truth about Peat Moss," *The Ecologist*, January 25, 2013, https:// theecologist.org/2013/jan/25/truth-about-peat-moss.

2. Michael Simone-Finstrom And Marla Spivak, "Propolis and Bee Health: The Natural History and Significance of Resin Use by Honey Bees," *Apidologie* 41, no. 3 (2010), https://doi.org/10.1051/apido/2010016.

3. Mark L. Winston, *Biology of the Honey Bee* (Cambridge, MA; London, England: Harvard University Press, 1991), 152.

4. Andrew B. Barron and Jenny Aino Plath, "The Evolution of Honey Bee Dance Communication: A Mechanistic Perspective," *Journal of Experimental Biology* 220 (2017), https://doi.org/10.1242/jeb.142778.

5. Josh Howgego, "Honeybees Gang up to Roast Invading Hornets Alive— at a Terrible Cost," *NewScientist*, July 16, 2018, https://www.newscientist .com/article/2174097-honeybees-gang-up-to-roast-invading-hornets-alive-at -a-terrible-cost/#ixzz6QEsxJNjw.

6. Harry Cockburn, "Bumblebees Face Mass Extinction Amid 'Climate Chaos,' Scientists Warn," *The Independent*, February 6, 2020, https://www.independent.co.uk/environment/bumblebee-extinction-climate-change-biodiversity-europe-america-a9321836.html?fbclid=IwAR1z8zxpgMKTXIrSBheavZnhfEe75K1X71CCcHtd4e5RNto6oUpLeyFKy2g.

7. Intangible Cultural Heritage, "Files 2020 under Process," *UNESCO*, accessed June 15, 2020, https://ich.unesco.org/en/files-2020-under-process-01053?select_country=00176&select_type=all.

8. George Monbiot, "Insectageddon: Farming Is More Catastrophic Than Climate Breakdown," *The Guardian*, October 20, 2017, https://www.theguardian.com/commentisfree/2017/oct/20/insectageddon-farming-catastrophe-climate-breakdown-insect-populations.

9. C. A. Hallmann et al., "More Than 75 Percent Decline over 27 Years in Total Flying Insect Biomass in Protected Areas," *PLoS ONE* 12, no. 10 (2017), https://doi.org/10.1371/journal.pone.0185809.

10. Welt, "Beekeepers Are Concerned about Pesticide Residues in Bee Pollen," November 18, 2019, https://www.welt.de/regionales/rheinland-pfalz-saarland/article203599438/Imker-besorgt-ueber-Rueckstaende-von-Pestiziden-in-Bienenpollen.html?fbclid=IwAR3c_q0dKY7L8wjY444qNT0wPZjBT5asRknDZ6r5noX9Ow9t5f7-4H80jQ0.

11. Theresa L. Pitts-Singer and Rosalind R. James, "Bees in Nature and on the Farm," in *Bee Pollination in Agricultural Ecosystems*, edited by Rosalind James, Rosalind R. James, Theresa L. Pitts-Singer (Oxford: Oxford University Press, 2008), 9.

12. Philip Donkersley, "Trees for Bees," *Agriculture, Ecosystems & Environment* 270–271 (February 2019), https://doi.org/10.1016/j.agee.2018.10.024.

13. Marina Koren, "Could Disappearing Wild Insects Trigger a Global Crop Crisis?" *Smithsonian Magazine*, February 28, 2013, https://www.smithsonianmag.com/science-nature/could-disappearing-wild-insects-trigger-a-global-crop-crisis-738/.

14. Adam Hart, "Huge Scale of California Pollination Event," *BBC*, March 26, 2013, https://www.bbc.com/news/science-environment-21741651.

15. Michael Greshko, "Oldest Evidence of Modern Bees Found in Argentina," *National Geographic*, February 11, 2020, https://www.nationalgeographic.com/science/2020/02/oldest-ever-fossil-bee-nests-discovered-in-patagonia/.

16. Patricia J. Campbell et al., *An Introduction to Global Studies* (Chichester: Wiley-Blackwell, 2010), 298.

17. EELP Staff, "Neonicotinoid Ban in National Wildlife Refuges," *Environmental and Energy Law Program (EELP)*, October 8, 2019, https://eelp.law.harvard.edu/2019/10/neonicotinoid-ban-in-national-wildlife-refuges/.americas/us-bee-pesticides-ban-overturned-neonic-population-decline-wildlife-refuges-a8477631.html.

18. Tom Barnes, "Trump Administration Reverses Ban on Bee-Harming Pesticides in Wildlife Refuges," *The Independent*, August 4, 2018, https://www.independent.co.uk/news/world/.

19. Joe Sandler Clarke, "Brazil Pesticide Approvals Soar as Jair Bolsonaro Moves to Weaken Rules," *Unearthed*, December 6, 2019, https://unearthed.greenpeace.org/2019/06/12/jair-bolsonaro-brazil-pesticides/.

20. E360 Digest, "Pesticides Linked to Deaths of Millions of Bees in Brazil," *Yale Environment 360*, August 29, 2019, https://e360.yale.edu/digest/pesticides-linked-to-deaths-of-millions-of-bees-in-brazil.

21. BBC News, "South African Bees: 'One Million Die in Cape Town,'" *BBC*, November 26, 2018, https://www.bbc.com/news/world-africa-46345127.

22. Department of Entomology, "Ground Nesting Bees in Your Backyard!" *Cornell CALS*, accessed June 24, 2020, https://entomology.cals.cornell.edu/extension/wild-pollinators/native-bees-your-backyard/.

23. Exodus 20:13, *The Bible* (Peabody, MA: Hendrickson Publishers, 2006), 400th anniversary edition, Kindle, 36.

24. Christian Schwägerl, "In Conservative Bavaria, Citizens Force Bold Action on Protecting Nature," *Yale Environment 360*, April 25, 2019, https://e360.yale.edu/features/in-conservative-bavaria-citizens-force-bold-action-on-protecting-nature.

25. Associated Press, "US Poultry Workers Wear Diapers on Job over Lack of Bathroom Breaks—Report," *The Guardian*, May 12, 2016, https://www.theguardian.com/us-news/2016/may/12/poultry-workers-wear-diapers-work-bathroom-breaks.

26. Murray Brewster, "The High-Speed Hard Sell: Why the F-35 Is Coming to a Canadian Air Show," *CBC*, September 5, 2019, https://www.cbc.ca/news/politics/f35-canada-competition-1.5270600.

CHAPTER 11: FLIGHT OF THE HONEYBEE

1. Henry Tumusiime, in discussion with the author, Bowa, Uganda, October 2019.

2. Tumusiime, in discussion with the author.

3. Tumusiime, in discussion with the author.

4. Tumusiime, in discussion with the author.

5. Tumusiime, in discussion with the author.

6. Simon Turner, in discussion with the author, multiple locations in Uganda, October 2019.

7. Turner, in discussion with the author.

8. Turner, in discussion with the author.

9. Turner, in discussion with the author.

10. Turner, in discussion with the author.

11. Tumusiime, in discussion with the author.

12. Simon Turner, *Beekeeping as a Business: A Practical Guide to Beekeeping in Uganda*, 4th ed. (Kampala, Uganda: Malaika Honey, 2018).

13. Kenyan Top Bar (KTB) hives are an alternative to traditional log hives. In a KTB hive, the colony is contained within a box with multiple bars that hold the honeycomb in place. In doing so, the beekeeper has easy access to the colony and can remove bars individually to inspect large numbers of bees and the honeycomb. The log hives, variations of which I have seen in both Uganda and Kenya (chapter 11), offer a simple, low-cost way to house a colony in a small hollowed-out trunk. The upside is the low or zero-capital cost to implement a log hive. However, a downside is that the small access area may agitate the bees during hive maintenance.

14. Turner, in discussion with the author.

15. Thomas Ayshford, in discussion with the author, Kampala, Uganda, October 2019.

16. Ayshford, in discussion with the author.

17. Ayshford, in discussion with the author.

18. Ayshford, in discussion with the author.

19. Ayshford, in discussion with the author.

20. Ayshford, in discussion with the author.

21. Ayshford, in discussion with the author.

22. DW, "Roundup Weedkiller: 42,000 Plaintiffs Sue Bayer over Glyphosate," October 30, 2019, https://www.dw.com/en/roundup-weedkiller-42000 -plaintiffs-sue-bayer-over-glyphosate/a-51043520.

23. DW, "Bayer Agrees $39.6 Million Settlement over Monsanto's Roundup Labeling," March 31, 2020, https://www.dw.com/en/bayer-agrees-396 -million-settlement-over-monsantos-roundup-labeling/a-52964239.

24. Eric Simons, "California's Early June Heat Wave Cooked Coastal Mussels in Place," *Bay Nature*, June 26, 2019, https://baynature.org/2019/06/26 /californias-early-june-heat-wave-cooked-coastal-mussels-in-place/.

CHAPTER 12: THANK YOU FOR THE RAIN

1. Julia Dahr, *Thank You for the Rain* (2017), https://www.imdb.com/title /tt6512856/?ref_=ttpl_pl_tt.

2. Kisilu Musya, in discussion with the author, multiple locations in Kenya, October 2019.

3. Matt McGrath, "Climate Change: Emissions Edge up Despite Drop in Coal," *BBC*, December 4, 2019, https://www.bbc.com/news/science-environ ment-50648495?SThisFB&fbclid=IwAR3z0oigZZE2mDwXV4VL2chzN0zz oUQmx2f3iJNr1jvpHUyZCUDmhENNd_4.

4. Kisilu Musya, in discussion with the author.

5. Musya, in discussion with the author.

6. Musya, in discussion with the author.

7. Musya, in discussion with the author.

8. Ibrahim Kioko Ngeke, in discussion with the author, Kitui county, Kenya, October 2019.

9. Musya, in discussion with the author.

10. Musya, in discussion with the author.

11. Ngeke, in discussion with the author.

12. Ngeke, in discussion with the author.

13. Musya, in discussion with the author.

14. Ngeke, in discussion with the author.

15. Musya, in discussion with the author.

16. Ngeke, in discussion with the author.

17. Pseudonym used.

18. Bob, in discussion with the author, Kitui County, Kenya, October 2019.

19. Bob, in discussion with the author.

20. Bob, in discussion with the author.

21. Bob, in discussion with the author.

22. Matthew L. Forister et al., "Declines in Insect Abundance and Diversity: We Know Enough to Act Now," *Conservation Science and Practice*, June 22, 2019, https://doi.org/10.1111/csp2.80.

23. Paul de Zylva, "Insectageddon—What's Happening to Bees and Other Insects?" *Friends of The Earth Policy*, April 30, 2019, https://policy.friendsof theearth.uk/opinion/insectageddon-whats-happening-bees-and-other-insects.

24. Wollheim Memorial, "Zyklon B: An Insecticide Becomes a Means for Mass Murder," accessed June 26, 2020, http://www.wollheim-memorial.de /en/zyklon_b_en_2.

25. Holocaust Encyclopedia, "Documenting Numbers of Victims of the Holocaust and Nazi Persecution," accessed June 26, 2020, https://encyclopedia .ushmm.org/content/en/article/documenting-numbers-of-victims-of-the -holocaust-and-nazi-persecution.

26. Andrew McGuire, "Farms Are Not Like Eden: The Case for Aggressive Human Intervention in Agriculture," *Genetic Literacy Project*, March 20, 2017, https://geneticliteracyproject.org/2017/03/20/farms-not-like-eden-case -aggressive-human-intervention-agriculture/.

27. Ian Johnston, "Farmer Suicides Soar in India as Deadly Heatwave Hits 51 Degrees Celsius," *The Independent*, May 20, 2016, https://www.independent.co.uk/news/world/asia/india-heatwave-farmers-suicide-killing-themselves-51-record-temperature-climate-change-global-a7039841.html.

28. Euronews, "Antarctica Iceberg: Huge Piece Breaks away from Ice Shelf," October 1, 2019, https://www.euronews.com/2019/09/30/massive-iceberg-breaks-away-from-antarctica-s-amery-ice-shelf.

29. IMDb, "Thank You for the Rain—Awards," accessed June 24, 2020, https://www.imdb.com/title/tt6512856/awards.

30. Musya, in discussion with the author.

31. Wanja Emily, written exchanges, 2019.

32. World Meteorological Organization, "WMO Confirms 2019 as Second Hottest Year on Record," January 15, 2020, https://public.wmo.int/en/media/press-release/wmo-confirms-2019-second-hottest-year-record.

33. Daniel Boffey, "EU Accused of Climate Crisis Hypocrisy after Backing 32 Gas Projects," *The Guardian*, February 12, 2020, https://amp.theguardian.com/environment/2020/feb/12/eu-accused-of-climate-crisis-hypocrisy-after-backing-32-gas-projects?__twitter_impression=true&fbclid=IwAR1aQDCyYG7Bea3F1zhP8lScMP0yqa_2GiCs8gB3Axlrfb4H016kUosOLrI.

34. Saul Elbein, "How to Live with Mega-Fires? Portugal's Feral Forests May Hold the Secret," *National Geographic*, December 6, 2019, https://www.nationalgeographic.com/science/2019/12/how-to-live-with-mega-fires-portugal-forests-may-hold-secret/.

35. Manuel Moncada, "Un 20 % del suelo en España está degradado y es 'muy vulnerable' ante la desertificación," *EFE Verde*, June 17, 2019, https://www.efeverde.com/noticias/20-del-suelo-espana-vulnerable-ante-desertificacion/.

36. Zoe Tidman, "A British Citizen Emits More CO2 in Two Weeks Than Some People in Africa Do in a Year, Research Shows," *The Independent*, January 6, 2020, https://www.independent.co.uk/environment/british-carbon-footprint-africa-emissions-oxfam-climate-change-a9271861.html.

37. World Nuclear Association, "Nuclear Power in France," accessed June 25, 2020, https://www.world-nuclear.org/information-library/country-profiles/countries-a-f/france.aspx.

38. Livemint, "France Inaugurates World's First Solar Highway," December 23, 2016, https://www.livemint.com/Politics/iEh2DTbnIcKUtVzGXr7LgP/France-inaugurates-worlds-first-solar-highway.html.

39. Joel Hruska, "France's Solar Road Is a Complete Failure," *Extreme-Tech*, August 20, 2019, https://www.extremetech.com/extreme/296951-frances-solar-road-is-a-complete-failure.

CHAPTER 13: JOURNEY SOUTH

1. Andrew Simpson, ed., *Language and National Identity in Africa* (Oxford: Oxford University Press, 2008), 296.

2. Pseudonym used to protect anonymity.

3. Pseudonym used to protect anonymity.

4. Pseudonym used to protect anonymity.

5. Columbus S. Mavhunga and Bukola Adebayo, "Zimbabwe's Water Woes Worsen as Capital Shuts Down Treatment Plant," *CNN*, September 4, 2019, https://edition.cnn.com/2019/09/24/africa/harare-water-crisis-intl/index.html.

6. Rutendo Mawere and Gibbs Dube, "Doctors' Strike Leaves Some Zimbabweans 'Gathering Here in Pain,'" *Voice of America*, October 4, 2019, https://www.voanews.com/africa/doctors-strike-leaves-some-zimbabweans-gathering-here-pain.

7. Lenin Ndebele, "Zim Soldiers Close to Starving," *Times Live*, November 10, 2019, https://www.timeslive.co.za/sunday-times/news/2019-11-10-zim-soldiers-close-to-starving/.

8. Associated Press, "Zimbabwe Says 200 Elephants Have Now Died amid Drought," *Voice of America*, November 12, 2019, https://www.voanews.com/africa/zimbabwe-says-200-elephants-have-now-died-amid-drought.

9. David Kenrick, *Decolonisation, Identity and Nation in Rhodesia, 1964–1979: A Race against Time* (London: Palgrave Macmillan, 2019), 143.

10. United Nations, Department of Economic and Social Affairs, Population Division, *World Population Prospects 2019* (United Nations, 2019). Online ed., rev. 1, https://population.un.org/wpp/Download/Files/1_Indicators%20(Standard)/EXCEL_FILES/1_Population/WPP2019_POP_F01_1_TOTAL_POPULATION_BOTH_SEXES.xlsx.

CHAPTER 14: GENERATIONS OF MINERS

1. Victor, in discussion with the author, Rustemberg, South Africa, November 2019.

2. Pseudonym used to protect anonymity.

3. Rock removal from within the mine to leave behind a cavity is called a stope.

4. Victor, in discussion with the author.

5. Victor, in discussion with the author.

6. Pseudonym used to protect anonymity.

7. Pseudonym used to protect anonymity.

8. William, in discussion with the author, Soweto, South Africa, November 2019.

9. William, in discussion with the author.

10. William, in discussion with the author.

11. Moeketsi Baloe, in discussion with the author, Rustenburg, South Africa, November 2019.

12. Baloe, in discussion with the author.

13. Baloe, in discussion with the author.

14. Samantha Libreri, "'Let Us Share The Land'—Land Ownership Dominates South African Election Debate," *RTE*, March 19, 2019, https://www.rte.ie/news/world/2019/0319/1037200-south-africa/.

15. Emmanuel K. Akyeampong and Henry Louis Gates Jr., eds., *Dictionary of African Biography*, vols. 1–6 (Oxford: Oxford University Press, 2012), 456.

16. Chris van Wyk, *Helen Joseph* (South Africe: Awareness Publishing, 2003).

17. Helen Rappaport and Marian Wright Edelman, *Encyclopedia of Women Social Reformers*, vol. 1 (Santa Barbara: ABC-CLIO, 2001), 345.

18. David Shukman, "Electric Car Future May Depend on Deep Sea Mining," *BBC*, November 13, 2019, https://www.bbc.com/news/science-environment-49759626.

19. Mike Ives, "Drive to Mine the Deep Sea Raises Concerns over Impacts," *Yale Environment 360*, October 20, 2014, https://e360.yale.edu/features/drive_to_mine_the_deep_sea_raises_concerns_over_impacts.

20. James Gordon, "Cobalt: The Dark Side of a Clean Future," *Raconteur*, June 4, 2019, https://www.raconteur.net/business-innovation/cobalt-mining-human-rights.

21. Michael Nest, *Coltan* (Cambridge: Polity, 2011), 89.

22. Akbarali Thobhani, *Western Sahara since 1975 under Moroccan Administration* (Lewiston, NY: Edwin Mellen Press, 2002), 190.

23. New Internationalist, "Hunger Myths," December 5, 1992, https://newint.org/features/1992/12/05/hunger.

24. International Criminal Court, "Darfur, Sudan," accessed June 24, 2020, https://www.icc-cpi.int/darfur.

25. Nina Strochlic, "Six Years Ago, Boko Haram Kidnapped 276 Schoolgirls. Where Are They Now?" *National Geographic*, accessed June 26, 2020, https://www.nationalgeographic.com/magazine/2020/03/six-years-ago-boko-haram-kidnapped-276-schoolgirls-where-are-they-now/.

26. World Health Organization, "Ebola in the Democratic Republic of the Congo," accessed June 26, 2020, https://www.who.int/emergencies/diseases/ebola/drc-2019.

27. United Nations Security Council Meeting, "Humanitarian Crisis in Democratic Republic of Congo Will Worsen Without Political Transition, End to Violence, Speakers Warn Security Council," March 19, 2018, https://www.un.org/press/en/2018/sc13253.doc.htm.

28. In 2012, the then king of Spain, Juan Carlos, went on an expensive elephant hunting trip in Botswana while subjects in his kingdom struggled in a recession. Giles Tremlett, "Spain's King Juan Carlos under Fire over Elephant Hunting Trip," *The Guardian*, April 15, 2012, https://www.theguardian.com/world/2012/apr/15/spain-king-juan-carlos-hunting.

29. United Nations, Department of Economic and Social Affairs, Population Division. *World Population Prospects 2019* (United Nations, 2019). Online ed., rev. 1. https://population.un.org/wpp/Download/Files/1_Indicators%20(Stadard)/EXCEL_FILES/1_Population/WPP2019_POP_F01_1_TOTAL_POPULATION_BOTH_SEXES.xlsx.

30. Courtney Faal, "The Partition of Africa," *Black Past*, February 21, 2009, https://www.blackpast.org/global-african-history/partition-africa/.

31. John Simpson, "Peru's 'Copper Mountain' in Chinese Hands," *BBC News*, June 17, 2018, http://news.bbc.co.uk/2/hi/americas/7460364.stm.

32. Patrick Wintour, "Emmanuel Macron: 'More Choice Would Mean Fewer Children in Africa,'" *The Guardian*, September 26, 2018, https://www.theguardian.com/global-development/2018/sep/26/education-family-planning-key-africa-future-emmanuel-macron-un-general-assembly.

33. Tony Paterson, "Ursula von der Leyen: Is This the Next Woman to Become Chancellor of Germany?" *The Independent*, December 15, 2013, https://www.independent.co.uk/news/world/europe/ursula-von-der-leyen-is-this-the-next-woman-to-become-chancellor-of-germany-9006358.html.

34. Michael Bergin, "Danes Tout Family Planning Aid to Africa to 'Limit Migration' to Europe," *FP*, July 12, 2017, https://foreignpolicy.com/2017/07/12/danes-tout-family-planning-aid-to-africa-to-limit-migration-to-europe/.

35. While carrying out research for *Music and Coexistence: A Journey across the World in Search of Musicians Making a Difference*, I saw different ways in which musicians could create peace and better understanding through their art form. Music education, in the form of a rock school in a divided city that brought Christian Serbs and Muslim ethnic Albanian Kosovars together; a mariachi camp hosted by the University of North Texas in Denton, Texas; a samba school in Brazil; and others, demonstrated the transformative power of education: it improved young people's lives.

36. NAACP, "Criminal Justice Fact Sheet," accessed June 25, 2020, https://www.naacp.org/criminal-justice-fact-sheet/.

37. BBC News, "George Floyd: What Happened in the Final Moments of His Life," *BBC*, May 30, 2020, https://www.bbc.com/news/world-us-canada-52861726.

CHAPTER 15: IN THE SHADOW OF THE TOWERS

1. The Telegraph, "14 of the World's Most Beautiful Fountains (Drought Permitting)—Dubai Fountain," July 26, 2017, https://www.telegraph.co.uk /travel/galleries/most-beautiful-fountains/dubai-fountain/.
2. Fred M. Shelley and Reagan Metz, *Geography of Trafficking: From Drug Smuggling to Modern-Day Slavery* (Santa Barbara: ABC-CLIO, 2017), 277.

CHAPTER 16: HONG KONG ISOLATION

1. Debarati Guha, "Opinion: The Death of Indian Democracy," *DW*, January 8, 2020, https://www.dw.com/en/opinion-the-death-of-indian-demo cracy/a-51932462.
2. When I checked the CDC rating in February 2020, Hong Kong was rated as level 1, however, since the CDC website doesn't appear to have an audit history of ratings, only the most recent rating can be referenced. Centers for Disease Control and Prevention,"COVID-19 in Hong Kong," accessed December 20, 2020, https://wwwnc.cdc.gov/travel/notices/covid-2/corona virus-hong-kong.
3. Level 3 was the maximum rating in February 2020. The CDC added a fourth level in November 2020 for COVID-19 called: "Very high level of COVID-19". Centers for Disease Control and Prevention, "How CDC Determines the Level for COVID-19 Travel Health Notices," November 21, 2020, https://www.cdc.gov/coronavirus/2019-ncov/travelers/how-level-is -determined.html#:~:text=This%20new%204%2Dlevel%20system,Moderate %20level%20of%20COVID%2D19.
4. When I checked the CDC rating in February 2020, China was rated as level 3, however, since the CDC website doesn't appear to have an audit history of ratings, only the most recent rating can be referenced. Centers for Disease Control and Prevention, "COVID-19 in China,"accessed December 20, 2020, https://wwwnc.cdc.gov/travel/notices/covid-1/novel-coronavirus-china.
5. Helen Regan, "Italy Announces Lockdown as Global Coronavirus Cases Surpass 105,000," *CNN*, March 8, 2020, https://edition.cnn.com/2020/03/08 /asia/coronavirus-covid-19-update-intl-hnk/index.html.
6. Phil Thomas, "Coronavirus: US Surpasses Italy for Most Recorded Deaths in the World with 20,254," *The Independent*, April 11, 2020, https:// www.independent.co.uk/news/world/americas/coronavirus-us-most-con firmed-deaths-latest-covid-19-a9460866.html.

7. Xiao Wu et al., "Exposure to Air Pollution and COVID-19 Mortality in the United States," Department of Biostatistics, Harvard T. H. Chan School of Public Health, April 5, 2020, https://projects.iq.harvard.edu/files/covid-pm/files/pm_and_covid_mortality.pdf.

8. Dr. Vandana Shiva, "Ecological Reflections on the Coronavirus," *Medium*, March 23, 2020, https://medium.com/post-growth-institute/ecological-reflections-on-the-coronavirus-93d50bbfe9db.

9. Shiva, "Ecological Reflections on the Coronavirus."

10. Denis Malvy et al., "Ebola Virus Disease," *The Lancet* 39, no. 10174 (March 2019): 937, https://doi.org/10.1016/S0140-6736(18)33132-5.

11. The Conversation, "Global Urbanization Created the Conditions for the Current Coronavirus Pandemic," June 18, 2020, https://theconversation.com/global-urbanization-created-the-conditions-for-the-current-coronavirus-pandemic-137738.

12. Jesus Olivero et al., "Recent Loss of Closed Forests Is Associated with Ebola Virus Disease Outbreaks," *Sci Rep* 7, no. 14291 (2017): 1, https://doi.org/10.1038/s41598-017-14727-9.

13. Catherine Kim, "It Took a Pandemic for Cities to Finally Address Homelessness," *Vox*, April 21, 2020, https://www.vox.com/2020/4/21/21227629/coronavirus-homeless-covid-19-las-vegas-san-francisco.

14. IATA, "2036 Forecast Reveals Air Passengers Will Nearly Double to 7.8 Billion," October 24, 2017, https://www.iata.org/en/pressroom/pr/2017-10-24-01/.

15. BBC News, "Branson's Virgin Atlantic in Virus Bailout Talks," *BBC*, April 26, 2020, https://www.bbc.com/news/business-52431290.

16. Hilary Osborne, "Virgin awarded almost £2bn of NHS contracts in the past five years," *The Guardian*, August 5, 2018, https://www.theguardian.com/society/2018/aug/05/virgin-awarded-almost-2bn-of-nhs-contracts-in-the-past-five-years.

17. BBC News, "Coronavirus: UK Interest Rates Cut to Lowest Level Ever," *BBC*, March 19, 2020, https://www.bbc.com/news/business-51962982.

18. Leonardo Martinez-Diaz and Giulia Christianson, "Quantitative Easing for Economic Recovery Must Consider Climate Change," *World Resources Institute*, May 11, 2020, https://www.wri.org/blog/2020/05/coronavirus-responsible-quantitative-easing.

19. The term "flatten the curve" refers to a reduction in the rate of new infections per day, aiming to reduce the burden on health-care providers and, in turn, avoid them being overwhelmed with coronavirus patients.

20. Associated Press in Phoenix, Arizona, "Arizona Man Dies after Attempting to Take Trump Coronavirus 'Cure,'" *The Guardian*, March 24, 2020, https://www.theguardian.com/world/2020/mar/24/coronavirus-cure-kills-man-after-trump-touts-chloroquine-phosphate.

21. Joshua Berlinger et al., "Excavators Are Digging Mass Graves in Northwest Brazil," *CNN*, April 22, 2020, https://edition.cnn.com/world/live-news/coronavirus-pandemic-04-22-20-intl/h_a8aba54c21b1ba72a20ff9a74acee4ff.

22. Andy Gregory, "New York Alone Now Has More Cases Than Any Country as Drone Footage Shows State Using Mass Graves," *The Independent*, April 10, 2020, https://www.independent.co.uk/news/world/americas/new-york-mass-graves-cases-deaths-hart-island-bronx-rikers-latest-a9459106.html.

23. When I checked the CDC rating in February 2020, Japan was rated as level 2, however, since the CDC website doesn't appear to have an audit history of ratings, only the most recent rating can be referenced. Centers for Disease Control and Prevention, "COVID-19 in Japan," accessed December 20, 2020, https://wwwnc.cdc.gov/travel/notices/covid-4/coronavirus-japan.

CHAPTER 17:FROM HIROSHIMA WITH LOVE

1. United Nations, *Kyoto Protocol to the United Nations Framework Convention on Climate Change* (New York: United Nations, 1998), https://unfccc.int/resource/docs/convkp/kpeng.pdf.

2. Flora Charner and Alessandra Castelli, "Chile Won't Host APEC and COP25 Summits, After Weeks of Street Protests," *CNN*, October 30, 2019, https://edition.cnn.com/2019/10/30/americas/chile-protests-apec-cop25-hosting-canceled-intl/index.html.

3. Soh Horie, in discussion with the author, Hiroshima, Japan, February 2020.

4. Horie, in discussion with the author.

5. Horie, in discussion with the author.

6. Horie, in discussion with the author.

7. Horie, in discussion with the author.

8. Located less than a half a mile from the hypocenter in a quiet neighborhood, the café serves food and drinks, and in addition to numerous events organized there, also hosts a regular meeting group for atomic bomb survivors to talk with members of the public about their experiences.

CHAPTER 18: LESSONS LEARNED

1. Matthew Stock, "Air Pollution Plunges in European Cities amid Coronavirus Lockdown: Satellite Data," *Reuters*, March 27, 2020, https://

www.reuters.com/article/us-health-coronavirus-europe-pollution/air-pollution
-plunges-in-european-cities-amid-coronavirus-lockdown-satellite-data-id
USKBN21E2UK.

2. Vivek Kumar and Maya Jamieson, "Himalayas Visible for First Time in 30 Years as Pollution Levels in India Drop," *SBS Hindi*, April 7, 2020, https://www.sbs.com.au/language/english/audio/himalayas-visible-for-first-time-in
-30-years-as-pollution-levels-in-india-drop?fbclid=IwAR3bTc8B0S3ykR36B
Rhul4tSXl1syuE9PrWzkZAKO5Y88qJx5Gldo-BBptA&cid=lang%3Asocial
share%3Afacebook.

3. BBC News, "Coronavirus: Wild Animals Enjoy Freedom of a Quieter World," *BBC*, April 29, 2020, https://www.bbc.com/news/world-52459487.

4. Alejandro Schwedhelm et al., "Biking Provides a Critical Lifeline during the Coronavirus Crisis," *World Resources Institute*, April 17, 2020, https://www
.wri.org/blog/2020/04/coronavirus-biking-critical-in-cities.

5. Louise Boyle, "The Ozone Layer Is Healing, New Study Finds," *The Independent*, March 27, 2020, https://www.independent.co.uk/environment
/ozone-layer-healing-coronavirus-climate-sun-uv-rays-a9429341.html.

6. Lisa Shumaker, "U.S. coronavirus deaths exceed 70,000 as forecasting models predict grim summer," *Reuters*, May 5, 2020, https://www.reuters
.com/article/us-health-coronavirus-usa-casualties-idUSKBN22H2ES.

7. BBC News, "Coronavirus: White House Plans to Disband Virus Task Force," *BBC*, May 6, 2020, https://www.bbc.com/news/world-us-canada
-52553829.

8. Samuel Lovett, "'We'd Never Seen a Temperature This High': 20C Mark Exceeded in Antarctica for First Time," *The Independent*, February 14, 2020, https://www.independent.co.uk/environment/climate-change/climate
-crisis-antarctica-temperature-record-seymour-island-a9335401.html.

9. David Rotman, "A Cheap and Easy Plan to Stop Global Warming," *MIT Technology Review*, February 8, 2013, https://www.technologyreview
.com/2013/02/08/84239/a-cheap-and-easy-plan-to-stop-global-warming/.

10. Alex Kirby, "Scientists Pour Cold Water on Ocean Geoengineering Idea," *Climate Home News*, January 28, 2016, https://www.climatechangenews
.com/2016/01/28/scientists-pour-cold-water-on-ocean-geoengineering-idea/.

11. Jesse L. Reynolds, *The Governance of Solar Geoengineering: Managing Climate Change in the Anthropocene* (Cambridge: Cambridge University Press, 2019), 23.

12. Kelsey Piper, "The Climate Renegade," *Vox*, June 4, 2019, https://www
.vox.com/the-highlight/2019/5/24/18273198/climate-change-russ-george
-unilateral-geoengineering.

13. M. G. Lawrence et al., "Evaluating Climate Geoengineering Proposals in the Context of the Paris Agreement Temperature Goals," *Nat Commun* 9, no. 3734 (2018): 13, https://doi.org/10.1038/s41467-018-05938-3.

14. Timothy M. Lenton, "Climate Tipping Points—Too Risky to Bet Against," *Nature*, November 27, 2019, https://www.nature.com/articles/d41586 -019-03595-0.

15. Sigal Samuel, "A Staggering 1 Billion Animals Are Now Estimated Dead in Australia's Fires," *Vox*, January 7, 2020, https://www.vox.com/future -perfect/2020/1/6/21051897/australia-fires-billion-animals-dead-estimate.

16. Christina Maxouris, "Australian PM Scott Morrison Says He Will 'Accept the Criticism' for Vacationing in Hawaii as Fires Raged Back Home," *CNN*, December 22, 2019, https://edition.cnn.com/2019/12/22/asia/australia -fire-prime-minister-criticism-apology/index.html.

17. Shaoni Bhattacharya, "The 2003 European Heatwave Caused 35,000 Deaths," *New Scientist*, October 10, 2003, https://www.newscientist.com/article /dn4259-the-2003-european-heatwave-caused-35000-deaths/#:~:text=At%20 least%2035%2C000%20people%20died,extreme%20weather%20events%20 lie%20ahead%E2%80%9D.

18. Hilary Whiteman, "India Heat Wave Kills 2,330 People as Millions Wait for Rain," *CNN*, June 2, 2015, https://edition.cnn.com/2015/06/01/asia /india-heat-wave-deaths/index.html.

19. Samuel Osborne, "France's Summer Heatwave Killed 1,500 People, Health Minister Announces," *The Independent*, September 8, 2019, https:// www.independent.co.uk/news/world/europe/france-heatwave-summer-dead -death-toll-europe-a9096806.html.

20. Samuel Petrequin, "EU Lays out 1 Trillion-Euro Plan to Support Green Deal," *AP News*, January 14, 2020, https://apnews.com/5d4db8ffda58f03f090 a04c35f0a2dc8.

21. Rachel Ehrenberg, "Global Count Reaches 3 Trillion Trees," *Nature*, September 2, 2015, https://www.nature.com/news/global-count-reaches-3 -trillion-trees-1.18287.

22. United Nations, Department of Economic and Social Affairs, *World Economic and Social Survey 2011: The Great Green Technological Transformation* (New York: United Nations, 2011), 27, https://www.un.org/development /desa/dpad/wp-content/uploads/sites/45/2011wess.pdf.

23. Rachel McGhee and staff, "Hundreds of Coal Jobs Lost as Low Coal Prices, Coronavirus Impacts Mining," *ABC*, June 18, 2020, https://www .abc.net.au/news/2020-06-19/hundreds-of-coal-jobs-lost-as-low-coal-prices -coronavirus-impact/12368804.

24. BP, "BP at a Glance," accessed June 16, 2020, https://www.bp.com/en /global/corporate/what-we-do/bp-at-a-glance.html.

25. Ben Chapman, "BP Sets Target to Achieve Net Zero by 2050 to Tackle Climate Emergency," *The Independent*, February 12, 2020, https://www.inde pendent.co.uk/news/business/news/bp-net-zero-target-2050-climate-change -emergency-a9332216.html.

26. Alice Azania-Jarvis, "BP Oil Spill: Disaster by Numbers," *The Independent*, September 14, 2010, https://www.independent.co.uk/environment /bp-oil-spill-disaster-by-numbers-2078396.html.

27. BP's annual report in 2010 to the United States Securities and Exchange Commission (SEC), the body responsible for the stock market, revealed that Iain Conn, chief executive of refining and marketing, took home a base salary of 690,000 British pounds, and that both he and Dr. Byron Grote, chief financial officer, whose base salary was $1.38 million per year, both received a 30 percent bonus because they had "met or exceeded their specific segment/ functional targets for the year." It is hard for any reasonable or sensible person to believe, or accept, that people responsible for the accident should have been given a bonus at all. BP, *Annual Report and Form 20-F 2010*, accessed June 18, 2020, 112, https://www.bp.com/content/dam/bp/business-sites/en/global /corporate/pdfs/investors/bp-annual-report-and-form-20f-2010.pdf.

28. Osseily Hanna, *Music and Coexistence: A Journey across the World in Search of Musicians Making a Difference* (Lanham, MD: Rowman & Littlefield, 2014), 52.

AFTERWORD

1. Erich Fromm, *To Have or To Be* (New York: Open Road Integrated Media, 2013), Kindle, 69.

2. Fromm, *To Have or To Be*, 87.

Bibliography

Akyeampong, Emmanuel K., and Henry Louis Gates Jr., eds. *Dictionary of African Biography*, vols. 1–6. Oxford: Oxford University Press, 2012.

Amnesty International. "The World's Refugees in Numbers," *Amnesty International*, accessed June 20, 2020, https://www.amnesty.org/en/what-we-do /refugees-asylum-seekers-and-migrants/global-refugee-crisis-statistics-and -facts/.

ANRed. "Otro ataque de Manaos a las comunidades campesinas en Santiago del Estero," July 5, 2016, https://www.anred.org/2016/07/05/otro-ataque -de-manaos-a-las-comunidades-campesinas-en-santiago-del-estero/.

Associated Press. "US Poultry Workers Wear Diapers on Job over Lack of Bathroom Breaks—Report," *The Guardian*, May 12, 2016, https://www.the guardian.com/us-news/2016/may/12/poultry-workers-wear-diapers-work -bathroom-breaks.

———. "Zimbabwe Says 200 Elephants Have Now Died amid Drought," *Voice of America*, November 12, 2019, https://www.voanews.com/africa /zimbabwe-says-200-elephants-have-now-died-amid-drought.

Associated Press in Phoenix, Arizona. "Arizona Man Dies after Attempting to Take Trump Coronavirus 'Cure,'" *The Guardian*, March 24, 2020, https:// www.theguardian.com/world/2020/mar/24/coronavirus-cure-kills-man -after-trump-touts-chloroquine-phosphate.

Azania-Jarvis, Alice. "BP Oil Spill: Disaster by Numbers," *The Independent*, September 14, 2010, https://www.independent.co.uk/environment/bp-oil -spill-disaster-by-numbers-2078396.html.

Barnes, Tom. "Trump Administration Reverses Ban on Bee-Harming Pesticides in Wildlife Refuges." *The Independent*, August 4, 2018, https://www.in dependent.co.uk/news/world/americas/us-bee-pesticides-ban-overturned -neonic-population-decline-wildlife-refuges-a8477631.html.

Barron, Andrew B., and Jenny Aino Plath, "The Evolution of Honey Bee Dance Communication: A Mechanistic Perspective," *Journal of Experimental Biology* 220 (2017): 4339–46, https://doi.org/10.1242/jeb.142778.

Bauman, Zygmunt. *Liquid Fear*. Cambridge: Polity Press, 2006.

BBC News. "Blogger Travels from Sheffield to Essex Via Berlin to Save Cash." *BBC*, January 28, 2016, https://www.bbc.com/news/uk-england-35424393.

———. "Branson's Virgin Atlantic in Virus Bailout Talks." *BBC*, April 26, 2020, https://www.bbc.com/news/business-52431290.

———. "Brazil Mensalao Trial: Former Chief of Staff Jailed." *BBC*, November 16, 2013, https://www.bbc.com/news/world-latin-america-24967116.

———. "Coronavirus: UK Interest Rates Cut to Lowest Level Ever." *BBC*, March 19, 2020, https://www.bbc.com/news/business-51962982.

———. "Coronavirus: White House Plans to Disband Virus Task Force." *BBC*, May 6, 2020, https://www.bbc.com/news/world-us-canada-52553829.

———. "Coronavirus: Wild Animals Enjoy Freedom of a Quieter World." *BBC*, April 29, 2020, https://www.bbc.com/news/world-52459487.

———. "Former VW Boss Charged over Diesel Emissions Scandal." *BBC*, April 15, 2019, https://www.bbc.com/news/47937141.

———. "George Floyd: What Happened in the Final Moments of his Life." *BBC*, May 30, 2020, https://www.bbc.com/news/world-us-canada-52861726.

———. "Istanbul Reina Attack Suspect Says Nightclub Was Chosen at Random." *BBC*, January 19, 2017, https://www.bbc.com/news/world-europe-38673154.

———. "Mitsubishi Motors Admits Using Wrong Tests since 1991." *BBC*, April 26, 2016, https://www.bbc.com/news/business-36137719.

———. "Panama Papers: What Happened Next?" *BBC News*, December 26, 2016, https://www.bbc.com/news/world-38319026.

———. "Prince Charles Takes Part in Saudi Arabian Sword Dance." *BBC*, February 19, 2014, https://www.bbc.com/news/uk-26252091.

———. "Renewable Energy Capacity Overtakes Coal." *BBC*, October 25, 2016, https://www.bbc.com/news/business-37767250.

———. "Solar Impulse Completes Historic Round-The-World Trip." *BBC*, July 26, 2016. https://www.bbc.com/news/science-environment-36890563.

———. "South African Bees: 'One Million Die in Cape Town.'" *BBC*, November 26, 2018, https://www.bbc.com/news/world-africa-46345127.

———. "Sweden Sees Rare Fall in Air Passengers, as Flight-Shaming Takes Off." *BBC*, January 10, 2020, https://www.bbc.com/news/world-europe-51067440.

BBC Sport. "Rio Olympics 2016: Refugee Olympic Team Competed as 'Equal Human Beings.'" *BBC*, August 21, 2016, https://www.bbc.com/sport/olympics/37037273.

Beaubien, Jason. "Tijuana Violence Likely to Continue in 2009." *NPR*, February 1, 2009, https://www.npr.org/templates/story/story.php?storyId=100063216.

Bergin, Michael. "Danes Tout Family Planning Aid to Africa to 'Limit Migration' to Europe." *FP*, July 12, 2017, https://foreignpolicy.com/2017/07/12 /danes-tout-family-planning-aid-to-africa-to-limit-migration-to-europe/.

Berlinger, Joshua, Adam Renton, Rob Picheta, and Rahim Zamira. "Excavators Are Digging Mass Graves in Northwest Brazil." *CNN*, April 22, 2020, https://edition.cnn.com/world/live-news/coronavirus-pandemic-04-22 -20-intl/h_a8aba54c21b1ba72a20ff9a74acee4ff.

Bhattacharya, Shaoni. "The 2003 European Heatwave Caused 35,000 Deaths." *New Scientist*, October 10, 2003, https://www.newscientist.com /article/dn4259-the-2003-european-heatwave-caused-35000-deaths /#:~:text=At%20least%2035%2C000%20people%20died,extreme%20 weather%20events%20lie%20ahead%E2%80%9D.

Boffey, Daniel. "EU Accused of Climate Crisis Hypocrisy after Backing 32 Gas Projects." *The Guardian*, February 12, 2020, https://amp.theguardian .com/environment/2020/feb/12/eu-accused-of-climate-crisis-hypocrisy -after-backing-32-gas-projects?__twitter_impression=true&fbclid=IwAR1a QDCyYG7Bea3F1zhP8lScMP0yqa_2GiCs8gB3Axlrfb4H016kUosOLrI.

Borunda, Alejandra. "Most Dire Projection of Sea-Level Rise Is a Little Less Likely, Reports Say." *National Geographic*, February 6, 2019, https://www .nationalgeographic.com/environment/2019/02/antarctic-greenland-ice -melt-less-bad/.

Bostock, Bill. "These 12 Charts Show How the World's Population Has Exploded in the Last 200 Years." *World Economic Forum*, July 15, 2019, https:// www.weforum.org/agenda/2019/07/populations-around-world-changed -over-the-years.

Boyle, Louise. "The Ozone Layer Is Healing, New Study Finds." *The Independent*, March 27, 2020, https://www.independent.co.uk/environment/ozone -layer-healing-coronavirus-climate-sun-uv-rays-a9429341.html.

BP. *Annual Report and Form 20-F 2010*, accessed June 18, 2020. https://www .bp.com/content/dam/bp/business-sites/en/global/corporate/pdfs/investors /bp-annual-report-and-form-20f-2010.pdf.

———. "BP at a Glance." Accessed June 16, 2020, https://www.bp.com/en /global/corporate/what-we-do/bp-at-a-glance.html.

Brewster, Murray. "The High-Speed Hard Sell: Why the F-35 Is Coming to a Canadian Air Show." *CBC*, September 5, 2019, https://www.cbc.ca/news /politics/f35-canada-competition-1.5270600.

Bulman, May. "Calais Jungle Closed: Hundreds of Children Remain Unaccounted for Despite Official Closure." *The Independent*, October 26, 2016, https://www.independent.co.uk/news/world/europe/calais-jungle-closed -children-unaccounted-official-closure-camp-refugees-a7381626.html.

Buncombe, Andrew. "US Tells UN It Is Officially Pulling out of Paris Agreement on Climate Change." *The Independent*, November 4, 2019, https://www.independent.co.uk/news/world/americas/us-politics/paris-agreement-climate-change-us-un-accord-pact-a9185171.html.

Buring, Andrea, and Alice Tidey. "Germany Offers Year of Rent to Asylum Seekers Who Return Home." *Euronews*, November 27, 2018, https://www.euronews.com/2018/11/27/germany-offers-year-of-rent-to-asylum-seekers-who-return-home.

Cachero, Paulina. "US Taxpayers Have Reportedly Paid an Average of $8,000 Each and Over $2 Trillion Total for the Iraq War Alone." *Business Insider*, February 6, 2020, https://www.businessinsider.com/us-taxpayers-spent-8000-each-2-trillion-iraq-war-study-2020-2?r=MX&IR=T.

Campbell, Patricia J., Aran MacKinnon, and Christy R. Stevens. *An Introduction to Global Studies*. Chichester: Wiley-Blackwell, 2010.

Canadian Press. "U.S. Methane Regulation Reversal Widens Its Competitive Advantage over Canada, Energy Industry Group Says." *CBC*, August 30, 2019, https://www.cbc.ca/news/canada/calgary/methane-regulations-canada-usa-alberta-oil-gas-capp-emissions-trump-1.5265988.

Centers for Disease Control and Prevention. "How CDC Determines the Level for COVID-19 Travel Health Notices." November 21, 2020, https://www.cdc.gov/coronavirus/2019-ncov/travelers/how-level-is-determined.html#:~:text=This%20new%204%2Dlevel%20system,Moderate%20level%20of%20COVID%2D19.

Centers for Disease Control and Prevention,"COVID-19 in Hong Kong." Accessed December 20, 2020, https://wwwnc.cdc.gov/travel/notices/covid-2/coronavirus-hong-kong.

Centers for Disease Control and Prevention, "COVID-19 in Japan." Accessed December 20, 2020, https://wwwnc.cdc.gov/travel/notices/covid-4/coronavirus-japan.

Centers for Disease Control and Prevention," COVID-19 in China." Accessed December 20, 2020, https://wwwnc.cdc.gov/travel/notices/covid-1/novel-coronavirus-china.

Chapman, Ben. "BP Sets Target to Achieve Net Zero by 2050 to Tackle Climate Emergency." *The Independent*, February 12, 2020, https://www.independent.co.uk/news/business/news/bp-net-zero-target-2050-climate-change-emergency-a9332216.html.

Charner, Flora, and Alessandra Castelli. "Chile Won't Host APEC and COP25 Summits, after Weeks of Street Protests." *CNN*, October 30, 2019, https://edition.cnn.com/2019/10/30/americas/chile-protests-apec-cop25-hosting-canceled-intl/index.html.

Chee, Foo Yun, and Kirstin Ridley. "EU Fines Barclays, Citi, JP Morgan, MUFG and RBS $1.2 Billion for FX Rigging." *Reuters*, May 16, 2019, https://

www.reuters.com/article/us-eu-antitrust-banks/eu-fines-barclays-citi -jp-morgan-mufg-and-rbs-12-billion-for-fx-rigging-idUSKCN1SM0XS.

City of Amsterdam. "Policy: Clean Air." Accessed June 25, 2020, https://www .amsterdam.nl/en/policy/sustainability/clean-air/.

Clarke, Joe Sandler. "Brazil Pesticide Approvals Soar as Jair Bolsonaro Moves to Weaken Rules." *Unearthed*, December 6, 2019, https://unearthed.green peace.org/2019/06/12/jair-bolsonaro-brazil-pesticides/.

Cockburn, Harry. "Bumblebees Face Mass Extinction amid 'Climate Chaos,' Scientists Warn." *The Independent*, February 6, 2020, https://www.indepen dent.co.uk/environment/bumblebee-extinction-climate-change-biodiversity -europe-america-a9321836.html?fbclid=IwAR1z8zxpgMKTXIrSBheavZn hfEe75K1X71CCcHtd4e5RNto6oUpLeyFKy2g.

Collingsworth, Terrence. "Class Complaint for Injunctive Relief and Damages." *International Rights Advocates*, December 15, 2019, http://iradvocates .org/sites/iradvocates.org/files/stamped%20-Complaint.pdf.

Columbia University. "EPA Revokes California's Authority to Set Climate-Protective Vehicle Emissions Standards." Accessed June 25, 2020, https:// climate.law.columbia.edu/content/epa-revokes-californias-authority-set -climate-protective-vehicle-emissions-standards.

———. "President Issues Executive Order Revoking Federal Sustainability Plan." Accessed June 25, 2020, https://climate.law.columbia.edu/content /president-issues-executive-order-revoking-federal-sustainability-plan-0.

The Conversation. "A Decade of Murder and Grief: Mexico's Drug War Turns Ten." December 11, 2016, https://theconversation.com/a-decade-of -murder-and-grief-mexicos-drug-war-turns-ten-70036.

———. "Global Urbanization Created the Conditions for the Current Coronavirus Pandemic." June 18, 2020, https://theconversation.com/global -urbanization-created-the-conditions-for-the-current-coronavirus-pan demic-137738.

Cowburn, Ashley. "David Cameron Offshore Fund: Video of Four Times PM Condemned 'Immoral' Tax Avoidance." *The Independent*, April 8, 2016, https://www.independent.co.uk/news/uk/politics/david-cameron-panama -papers-tax-avoidance-father-offshore-investment-fund-ian-cameron-blair more-a6974546.html.

Cowie, Sam. "The Jungle Metropolis: How Sprawling Manaus Is Eating into the Amazon." *The Guardian*, July 23, 2019, https://www.theguardian.com /cities/2019/jul/23/the-jungle-metropolis-how-sprawling-manaus-is-eating -into-the-amazon.

Cox, David. *Dirty Secrets, Dirty War: Buenos Aires, Argentina, 1976–1983: The Exile of Robert J. Cox*. Charleston: Evening Post, 2008.

Crutzen, Paul J., and Eugene F. Stoerner. "The 'Anthropocene.'" *Global Change Newsletter* 41 (2020): 17–18.

Cusick, Daniel. "Miami Is the "Most Vulnerable" Coastal City Worldwide." *Scientific American*, February 4, 2020, https://www.scientificamerican.com /article/miami-is-the-most-vulnerable-coastal-city-worldwide/.

Dahr, Julia. *Thank You for the Rain*. 2017, https://www.imdb.com/title /tt6512856/?ref_=ttpl_pl_tt.

Daugherty, Susan. "Tristram Stuart: Waging War against Global Food Waste." *National Geographic*, October 16, 2014, https://www.nationalgeographic .com/news/2014/10/141016-food-waste-tristram-stuart-emerging-ex plorer-hunger-charity-ngfood/.

Day, Michael, "Costa Concordia trial: Was Captain Francesco Schettino Re-ally the Only One at Fault for The Disaster?," *The Independent*, February 15, 2015, https://www.independent.co.uk/news/world/europe/costa-concordia -trial-was-captain-francesco-schettino-really-only-one-fault-disaster -10046725.html.

Dearden, Lizzie. "Billions of Pounds Worth of Weapons Have Been Licensed by the UK to Saudi Arabia since Start of Yemen War." *The Independent*, August 21, 2019, https://www.independent.co.uk/news/uk/home-news /saudi-arabia-arms-sales-uk-trade-yemen-war-crimes-a9073836.html.

de Castro, Nicolle Monteiro. "Brazil Center-South Cumulative Ethanol Production in Crop Year 2019–20 Rises 6.60% Year on Year: UNICA." *S&P Global*, January 27, 2020, https://www.spglobal.com/platts/en/market -insights/latest-news/agriculture/012720-brazil-center-south-cumulative -ethanol-production-in-crop-year-2019-20-rises-660-year-on-year-unica.

Department of Entomology. "Ground Nesting Bees in Your Backyard!" Cornell CALS, accessed June 24, 2020, https://entomology.cals.cornell.edu/exten sion/wild-pollinators/native-bees-your-backyard/.

Deutch, John M. *The Crisis in Energy Policy*. Cambridge, MA; London, Eng-land: Harvard University Press, 2011.

Deutsche Welle (DW). "Tax Authorities in Denmark Buy 'Panama Papers' Ev-idence." *DW*, September 29, 2016, https://www.dw.com/en/tax-authorities -in-denmark-buy-panama-papers-evidence/a-35928320.

de Zylva, Paul. "Insectageddon—What's Happening to Bees and Other Insects?" *Friends of the Earth Policy*, April 30, 2019, https://policy.friends oftheearth.uk/opinion/insectageddon-whats-happening-bees-and-other -insects.

Dias de Oliveira, Marcelo E., Burton E. Vaughan, and Edward J. Rykiel, "Eth-anol as Fuel: Energy, Carbon Dioxide Balances, and Ecological Footprint," *BioScience* 55, no. 7 (July 2005): 593–602, https://doi.org/10.1641/0006 -3568(2005)055[0593:EAFECD]2.0.CO;2.

DinoBuzz. "What Killed the Dinosaurs?" UCMP Berkeley, accessed June 20, 2020, https://ucmp.berkeley.edu/diapsids/extinctheory.html.

Donkersley, Philip, "Trees for Bees," *Agriculture, Ecosystems & Environment* 270–271 (February 2019): 79–83, https://doi.org/10.1016/j.agee.2018.10.024.

Douglas, Bruce. "Anger Rises as Brazilian Mine Disaster Threatens River and Sea with Toxic Mud." *The Guardian*, November 22, 2015, https://www.theguardian.com/business/2015/nov/22/anger-rises-as-brazilian-mine-disaster-threatens-river-and-sea-with-toxic-mud.

Dubnow, Shoshana. "Report Indicates Increase of Cocaine Usage in EU, Links to Social Media." *Euronews*, June 6, 2019, https://www.euronews.com/2019/06/06/report-indicates-increase-of-cocaine-usage-in-eu-links-to-social-media.

DW. "Bayer Agrees $39.6 Million Settlement over Monsanto's Roundup Labeling." March 31, 2020, https://www.dw.com/en/bayer-agrees-396-million-settlement-over-monsantos-roundup-labeling/a-52964239.

———. "Roundup Weedkiller: 42,000 Plaintiffs Sue Bayer over Glyphosate." October 30, 2019, https://www.dw.com/en/roundup-weedkiller-42000-plaintiffs-sue-bayer-over-glyphosate/a-51043520.

E360 Digest. "Pesticides Linked to Deaths of Millions of Bees in Brazil." *Yale Environment 360*, August 29, 2019, https://e360.yale.edu/digest/pesticides-linked-to-deaths-of-millions-of-bees-in-brazil.

EELP Staff. "Neonicotinoid Ban in National Wildlife Refuges." *Environmental and Energy Law Program (EELP)*, October 8, 2019, https://eelp.law.harvard.edu/2019/10/neonicotinoid-ban-in-national-wildlife-refuges/.

Ehrenberg, Rachel. "Global Count Reaches 3 Trillion Trees." *Nature*, September 2, 2015, https://www.nature.com/news/global-count-reaches-3-trillion-trees-1.18287.

Ehrlich, Paul R. *The Population Bomb*. San Francisco: Sierra Club/Ballantine Books, 1968.

Ehrlich, Paul R., and Anne H. Ehrlich. *The Population Explosion*. New York: Simon & Schuster, 1990.

Elbagir, Nima, Alison Main, Salma Abdelaziz, Laura Smith-Spark, and Jennifer Hansler. "The US Cleared the Way for a New Arms Sale to the UAE, Despite Evidence It Violated the Last One." *CNN*, May 22, 2020, https://edition.cnn.com/2020/05/22/world/state-department-uae-arms-sale-yemen-intl/index.html.

Elbein, Saul. "How to Live with Mega-Fires? Portugal's Feral Forests May Hold the Secret." *National Geographic*, December 6, 2019, https://www.nationalgeographic.com/science/2019/12/how-to-live-with-mega-fires-portugal-forests-may-hold-secret/.

Elgot, Jessica. "Nigel Farage: two of my children have German passports." *The Guardian*, April 18, 2018, https://www.theguardian.com/politics/2018/apr/18/nigel-farage-admits-his-children-hold-german-as-well-as-uk-passports.

Eshel, Gidon, Alon Shepon, Tamar Makov, and Ron Milo, "Land, Irrigation Water, Greenhouse Gas, and Reactive Nitrogen Burdens of Meat, Eggs, and Dairy Production in the United States," *Proceedings of the National Academy of Sciences of the United States of America* 111, no. 33 (Summer 2014), https://doi.org/10.1073/pnas.1402183111.

Eurofoodbank. "The Global Context of Food Waste." Accessed June 23, 2020, https://www.eurofoodbank.org/en/food-waste.

Euronews. "Antarctica Iceberg: Huge Piece Breaks away from Ice Shelf." October 1, 2019, https://www.euronews.com/2019/09/30/massive-iceberg-breaks-away-from-antarctica-s-amery-ice-shelf.

———. "'Ryanair Is the New Coal' as It Becomes First Airline in EU's Top Ten Biggest Polluters." April 3, 2019, https://www.euronews.com/2019/04/02/ryanair-is-the-new-coal-airline-is-eu-s-10th-biggest-polluter.

European Monitoring Centre for Drugs and Drug Addiction (EMCDDA). "Drug Use Prevalence and Trends." Accessed June 25, 2020, https://www.emcdda.europa.eu/publications/edr/trends-developments/2017/html/prevalence-trends/cocaine_en.

Exodus 20:13. *The Bible*. 400th anniversary ed. Peabody, MA: Hendrickson Publishers, 2006. Kindle.

Faal, Courtney. "The Partition of Africa." *Black Past*, February 21, 2009, https://www.blackpast.org/global-african-history/partition-africa/.

Feedback. "Feeding the 5,000." Accessed June 23, 2020, https://feedbackglobal.org/campaigns/feeding-the-5000/.

Forister, Matthew L., Emma M. Pelton, and Scott H. Black, "Declines in Insect Abundance and Diversity: We Know Enough to Act Now," *Conservation Science and Practice*, June 22, 2019, https://doi.org/10.1111/csp2.80.

Forster, Katie. "Paris Bans Cars with Even Plates in Third Day of Transport Restrictions to Ease Air Pollution." *The Independent*, December 8, 2016, https://www.independent.co.uk/news/world/europe/paris-banned-cars-even-number-plates-pollution-latest-public-transport-air-smog-bans-a7462621.html.

Fromm, Erich. *The Art of Loving*. New York: HarperCollins, 1956.

———. *To Have or To Be*. New York: Open Road Integrated Media, 2013. Kindle.

Gagne, David. "How Drug Traffickers Operate in Peru's Amazon Region." *InSight Crime*, April 7, 2015, https://www.insightcrime.org/news/brief/how-drug-traffickers-operate-peru-amazon/.

Galeano, Eduardo. *Open Veins of Latin America: Five Centuries of the Pillage of a Continent*. Translated by Cedric Belfrage. New York: Monthly Review Press, 1997. Kindle.

Gallucci, Maria. "At Last, the Shipping Industry Begins Cleaning up Its Dirty Fuels." *Yale Environment 360*, June 28, 2018, https://e360.yale.edu/features/at-last-the-shipping-industry-begins-cleaning-up-its-dirty-fuels.

Gamarro, Urías. "Guatemaltecos enviaron remesas por más de US$1 millón por hora en 2019." *Prensa Libre*, January 9, 2020, https://www.prensalibre.com/economia/guatemaltecos-enviaron-remesas-por-mas-de-us1-millon-por-hora-en-2019/.

Garside, Juliette. "Panama Papers: European Parliament Opens Inquiry." *The Guardian*, September 27, 2016, https://www.theguardian.com/world/2016/sep/27/panama-papers-inquiry-opens-at-european-parliament.

Global Witness. "On Dangerous Ground." June 20, 2016, https://www.globalwitness.org/en/campaigns/environmental-activists/dangerous-ground/.

Goldman Environmental Prize. "Isidro Baldenegro, 2005 Goldman Prize Recipient, North America." Accessed June 22, 2020, https://www.goldmanprize.org/recipient/isidro-baldenegro/.

Golson, Jordan. "The New Tesla Model S P100D Is Faster Than All These Cars." *The Verge*, August 23, 2016, https://www.theverge.com/2016/8/23/12615304/tesla-model-s-p100d-fastest-acceleration.

Gordon, James. "Cobalt: The Dark Side of a Clean Future." *Raconteur*, June 4, 2019, https://www.raconteur.net/business-innovation/cobalt-mining-human-rights.

Gov.uk. "UK Becomes First Major Economy to Pass Net Zero Emissions Law." June 27, 2019, https://www.gov.uk/government/news/uk-becomes-first-major-economy-to-pass-net-zero-emissions-law#:~:text=The%20UK%20today%20became%20the,80%25%20reduction%20from%201990%20levels.

Greenwalt, Megan. "A Look at the Largest Landfill Gas-To-Energy Project in Georgia." July 27, 2016, https://www.waste360.com/gas-energy/look-largest-landfill-gas-energy-project-georgia.

Gregory, Andy. "New York Alone Now Has More Cases Than Any Country as Drone Footage Shows State Using Mass Graves." *The Independent*, April 10, 2020, https://www.independent.co.uk/news/world/americas/new-york-mass-graves-cases-deaths-hart-island-bronx-rikers-latest-a9459106.html.

Greshko, Michael. "Oldest Evidence of Modern Bees Found in Argentina." *National Geographic*, February 11, 2020, https://www.nationalgeographic.com/science/2020/02/oldest-ever-fossil-bee-nests-discovered-in-patagonia/.

Grillo, Ioan. "Inside El Salvador's 'War without Sense.'" *Time*, July 23, 2015, https://time.com/3966900/el-salvador-gangs-violence/.

Guha, Debarati. "Opinion: The Death of Indian Democracy." *DW*, January 8, 2020, https://www.dw.com/en/opinion-the-death-of-indian-democracy/a-51932462.

Hallmann, C. A., M. Sorg, E. Jongejans, H. Siepel, N. Hofland, and H. Schwan, "More Than 75 Percent Decline over 27 Years in Total Flying

Insect Biomass in Protected Areas," *PLoS ONE* 12, no. 10 (2017), https:// doi.org/10.1371/journal.pone.0185809.

Hanna, Osseily. *Music and Coexistence: A Journey across the World in Search of Musicians Making a Difference.* Lanham, MD: Rowman & Littlefield, 2014.

Harding, Luke. "Panama Papers Investigation Wins Pulitzer Prize." *The Guardian*, April 11, 2017, https://www.theguardian.com/world/2017 /apr/11/panama-papers-investigation-wins-pulitzer-prize.

Harris, Gardiner. "Rex Tillerson Is Confirmed as Secretary of State Amid Record Opposition." *New York Times*, February 1, 2017, https://www.nytimes .com/2017/02/01/us/politics/rex-tillerson-secretary-of-state-confirmed .html.

Hart, Adam. "Huge Scale of California Pollination Event." *BBC*, March 26, 2013, https://www.bbc.com/news/science-environment-21741651.

Hemingway, Ernest. *To Have and Have Not.* New York: Scribner, 1937.

History. "The Pilgrims." November 21, 2019, https://www.history.com/topics /colonial-america/pilgrims.

Holocaust Encyclopedia. "Documenting Numbers of Victims of the Holocaust and Nazi Persecution." Accessed June, 26, 2020, https://encyclopedia .ushmm.org/content/en/article/documenting-numbers-of-victims-of-the -holocaust-and-nazi-persecution.

Hosie, Rachel. "What's It Like to Go on Holiday to Turkey Right Now?" *The Independent*, August 25, 2017, https://www.independent.co.uk/travel /europe/turkey-holidays-whats-it-like-bodrum-terror-attack-political -coup-summer-tourism-a7908636.html.

Howgego, Josh. "Honeybees Gang up to Roast Invading Hornets Alive—At a Terrible Cost." *NewScientist*, July 16, 2018, https://www.newscientist.com /article/2174097-honeybees-gang-up-to-roast-invading-hornets-alive-at-a -terrible-cost/#ixzz6QEsxJNjw.

Hruska, Joel. "France's Solar Road Is a Complete Failure." *ExtremeTech*, August 20, 2019, https://www.extremetech.com/extreme/296951-frances -solar-road-is-a-complete-failure.

Human Rights Watch. "Off the Radar." October 18, 2014, https://www.hrw .org/report/2014/10/18/radar/human-rights-tindouf-refugee-camps.

IAEA. "Country Nuclear Power Profiles: Turkey." Accessed June 25, 2020, https://www-pub.iaea.org/MTCD/Publications/PDF/cnpp2018/country profiles/Turkey/Turkey.htm.

IATA. "2036 Forecast Reveals Air Passengers Will Nearly Double to 7.8 Billion." October 24, 2017, https://www.iata.org/en/pressroom/pr/2017 -10-24-01/.

IEA. *Solar Energy Perspectives.* Paris: IEA, 2011, https://www.iea.org/reports /solar-energy-perspectives.

IIRSA. "Taller Sobre Agendas Cartográficas." Buenos Aires, Argentina, November 11, 2011, http://www.iirsa.org/admin_iirsa_web/Uploads/Docu ments/geo11_baires11_agenda_cartografica.pdf.

IMDb. "The Salt of the Earth." 2014. Accessed June 22, 2020, https://www .imdb.com/title/tt3674140/?ref_=ttmc_mc_tt.

———. "Thank You for the Rain—Awards." Accessed June 24, 2020, https:// www.imdb.com/title/tt6512856/awards.

Intangible Cultural Heritage. "Files 2020 under Process." UNESCO. Accessed June 15, 2020, https://ich.unesco.org/en/files-2020-under-process-01053 ?select_country=00176&select_type=all.

International Criminal Court. "Darfur, Sudan." Accessed June 24, 2020, https://www.icc-cpi.int/darfur.

ITS International. "Barcelona's Bike Share Scheme a Lifesaver." January 26, 2012, https://www.itsinternational.com/feature/barcelonas-bike-share -scheme-life-saver.

Ives, Mike. "Drive to Mine the Deep Sea Raises Concerns over Impacts." *Yale Environment 360*, October 20, 2014, https://e360.yale.edu/features /drive_to_mine_the_deep_sea_raises_concerns_over_impacts.

La Izquierda Diario. "Santiago del Estero: brutal ataque contra la Comunidad Bajo Fondo por parte de la empresa Manaos." September 24, 2016, http:// www.laizquierdadiario.com/Santiago-del-Estero-brutal-ataque-contra-la -Comunidad-Bajo-Fondo-por-parte-de-la-empresa-Manaos.

Jaitman, Laura, and Stephen Machin, "Crime and Immigration: New Evidence from England and Wales," *IZA J Migration* 2, no. 19 (2013), https://doi .org/10.1186/2193-9039-2-19.

Johnston, Ian. "Farmer Suicides Soar in India as Deadly Heatwave Hits 51 Degrees Celsius." *The Independent*, May 20, 2016, https://www.independent.co .uk/news/world/asia/india-heatwave-farmers-suicide-killing-themselves -51-record-temperature-climate-change-global-a7039841.html.

Jones, Nicola. "How the World Passed a Carbon Threshold and Why It Matters." *Yale Environment 360*, January 26, 2017, https://e360.yale.edu /features/how-the-world-passed-a-carbon-threshold-400ppm-and-why-it -matters.

Karp, Paul. "Coalition Says Panama Papers Sparked Action against More Than 100 Australian Taxpayers." *The Guardian*, September 6, 2016, https:// www.theguardian.com/australia-news/2016/sep/06/coalition-says-panama -papers-sparked-action-against-more-than-100-australian-taxpayers.

Kennedy, Dennis. "Environmental Degradation and Disrupted Social Fabric in the Tar Creek Basin." PhD thesis: Oklahoma State University, 2008.

Kenrick, David. *Decolonisation, Identity and Nation in Rhodesia, 1964–1979: A Race against Time*. London: Palgrave Macmillan, 2019.

Kim, Catherine. "It Took a Pandemic for Cities to Finally Address Homelessness." *Vox*, April 21, 2020, https://www.vox.com/2020/4/21/21227629/coronavirus-homeless-covid-19-las-vegas-san-francisco.

Kirby, Alex. "Scientists Pour Cold Water on Ocean Geoengineering Idea." *Climate Home News*, January 28, 2016, https://www.climatechangenews.com/2016/01/28/scientists-pour-cold-water-on-ocean-geoengineering-idea/.

Kirchgaessner, Stephanie. "Italy Tackles Food Waste with Law Encouraging Firms to Donate Food." *The Guardian*, August 3, 2016, https://www.theguardian.com/world/2016/aug/03/italy-food-waste-law-donate-food.

Kolbert, Elizabeth. *The Sixth Extinction: An Unnatural History*. London: Bloomsbury, 2014. Kindle.

Koren, Marina. "Could Disappearing Wild Insects Trigger a Global Crop Crisis?" *Smithsonian Magazine*, February 28, 2013, https://www.smithsonianmag.com/science-nature/could-disappearing-wild-insects-trigger-a-global-crop-crisis-738/.

Kottasová, Ivana. "Norway Says Its New Giant Oil Field Is Actually Good for the Environment. Critics Call It Climate Hypocrisy." *CNN*, January 19, 2020, https://amp.cnn.com/cnn/2020/01/19/business/norway-oil-field-climate-change-intl/index.html.

Kumar, Vivek, and Maya Jamieson. "Himalayas Visible for First Time in 30 Years as Pollution Levels in India Drop." *SBS Hindi*, April 7, 2020, https://www.sbs.com.au/language/english/audio/himalayas-visible-for-first-time-in-30-years-as-pollution-levels-in-india-drop?fbclid=IwAR3bTc8B0S3ykR36BRhul4tSXl1syuE9PrWzkZAKO5Y88qJx5G1do-BBptA&cid=lang%3Asocialshare%3Afacebook.

Lawrence, M. G., S. Schäfer, and H. Muri et al., "Evaluating Climate Geoengineering Proposals in the Context of the Paris Agreement Temperature Goals," *Nat Commun* 9, no. 3734 (2018), https://doi.org/10.1038/s41467-018-05938-3.

Lenton, Timothy M. "Climate Tipping Points—Too Risky to Bet Against." *Nature*, November 27, 2019, https://www.nature.com/articles/d41586-019-03595-0.

Libreri, Samantha. "'Let Us Share the Land'—Land Ownership Dominates South African Election Debate." *RTE*, March 19, 2019, https://www.rte.ie/news/world/2019/0319/1037200-south-africa/.

Lin, D., L. Hanscom, A. Murthy, A. Galli, M. Evans, E. Neill, M. S. Mancini, J. Martindill, F.-Z. Medouar, S. Huang, and M. Wackernagel, "Ecological Footprint Accounting for Countries: Updates and Results of the National Footprint Accounts, 2012–2018," *Resources* 7, no. 58, 1, https://www.mdpi.com/2079-9276/7/3/58#cite.

Livemint. "France Inaugurates World's First Solar Highway." December 23, 2016, https://www.livemint.com/Politics/iEh2DTbnIcKUtVzGXr7LgP /France-inaugurates-worlds-first-solar-highway.html.

Lo, Andrea. "The Netherlands Is Paying People to Cycle." *CNN*, December 21, 2018, https://edition.cnn.com/travel/article/netherlands-cycling/index .html.

Lovett, Samuel. "'We'd Never Seen a Temperature This High': 20C Mark Exceeded in Antarctica for First Time." *The Independent*, February 14, 2020, https://www.independent.co.uk/environment/climate-change/climate-crisis -antarctica-temperature-record-seymour-island-a9335401.html.

Lynn, Paul A. *Onshore and Offshore Wind Energy: An Introduction*. Chichester: Wiley, 2012.

Malthus, Thomas. *An Essay on the Principle of Population*. Digireads.com, 2013. Kindle.

Malvy, Denis, Anita K. McElroy, Hilde de Clerck, Stephan Günther, and Johan van Griensven, "Ebola Virus Disease," *The Lancet* 39, no. 10174 (March 2019): 936–48, https://doi.org/10.1016/S0140-6736(18)33132-5.

Mano, Ana. "Brazil's Beef Export Hit Record, Prospects Bright on China Demand." *Reuters*, December 10, 2019, https://www.reuters.com/article/us -brazil-beef/brazils-beef-export-hit-record-prospects-bright-on-china-de mand-idUSKBN1YE1TS.

Martinez-Diaz, Leonardo, and Giulia Christianson. "Quantitative Easing for Economic Recovery Must Consider Climate Change." *World Resources Institute*, May 11, 2020, https://www.wri.org/blog/2020/05/coronavirus -responsible-quantitative-easing.

Mavhunga, Columbus S., and Bukola Adebayo. "Zimbabwe's Water Woes Worsen as Capital Shuts down Treatment Plant." *CNN*, September 4, 2019, https://edition.cnn.com/2019/09/24/africa/harare-water-crisis-intl /index.html.

Mawere, Rutendo, and Gibbs Dube. "Doctors' Strike Leaves Some Zimbabweans 'Gathering Here in Pain.'" *VOA*, October 4, 2019, https://www.voanews .com/africa/doctors-strike-leaves-some-zimbabweans-gathering-here-pain.

Maxouris, Christina. "Australian PM Scott Morrison Says He Will 'Accept the Criticism' for Vacationing in Hawaii as Fires Raged Back Home." *CNN*, December 22, 2019, https://edition.cnn.com/2019/12/22/asia/australia -fire-prime-minister-criticism-apology/index.html.

McBride, James. "Understanding the Libor Scandal." Council on Foreign Relations, October 12, 2016, https://www.cfr.org/backgrounder/understanding -libor-scandal.

McGhee, Rachel, and staff. "Hundreds of Coal Jobs Lost as Low Coal Prices, Coronavirus Impacts Mining." *ABC*, June 18, 2020, https://www.abc.net

.au/news/2020-06-19/hundreds-of-coal-jobs-lost-as-low-coal-prices-coro navirus-impact/12368804.

McGrath, Matt. "Climate Change: Emissions Edge up Despite Drop in Coal." *BBC*, December 4, 2019, https://www.bbc.com/news/science-environment -50648495?SThisFB&fbclid=IwAR3z0oigZZE2mDwXV4VL2chzN0zzo UQmx2f3iJNr1jvpHUyZCUDmhENNd_4.

McGuire, Andrew. "Farms Are Not Like Eden: The Case for Aggressive Human Intervention in Agriculture." *Genetic Literacy Project*, March 20, 2017, https://geneticliteracyproject.org/2017/03/20/farms-not-like-eden-case -aggressive-human-intervention-agriculture/.

Mernit, Judith Lewis. "How Eating Seaweed Can Help Cows to Belch Less Methane." *Yale Environment 360*, July 2, 2018, https://e360.yale.edu/features /how-eating-seaweed-can-help-cows-to-belch-less-methane.

Merrill, Jamie. "Ugly Fruit Gets a Makeover as Store Cuts Waste," *The Independent*, May 31, 2014, https://www.independent.co.uk/life-style/food-and -drink/news/ugly-fruit-gets-makeover-store-cuts-waste-9466943.html.

Michelin Guide. "Michelin Guide Denmark 2020: Relae." Accessed June 22, 2020, https://guide.michelin.com/en/capital-region/kobenhavn/restaurant /relae.

Michigan Urban Farming Initiative. "What's Going on at MUFI?" Accessed June 23, 2020, https://www.miufi.org/projects.

Middelgrunden. "About Middelgrunden Wind Cooperative." Accessed June 25, 2020, http://www.middelgrunden.dk/middelgrunden/?q=en/node/35.

Ministerio da Saude, "Risk Assessment—Olympic and Paralympic Games." Brazil's Ministry of Health, 2016, http://www.saude.gov.br/images/pdf /2016/agosto/05/MA---2.pdf.

Mobike. "What is Mobike." Accessed June 25, 2020, https://mobike.com /global/faq.

Monbiot, George. "Insectageddon: Farming Is More Catastrophic Than Climate Breakdown." *The Guardian*, October 20, 2017, https://www.theguard ian.com/commentisfree/2017/oct/20/insectageddon-farming-catastrophe -climate-breakdown-insect-populations.

Moncada, Manuel. "Un 20 % del suelo en España está degradado y es 'muy vulnerable' ante la desertificación." *EFE Verde*, June 17, 2019, https://www .efeverde.com/noticias/20-del-suelo-espana-vulnerable-ante-desertificacion/.

Morales, Alberto. "Calderon and Pena Nieto Spent $675 Million on Planes and Helicopters." *El Universal*, November 24, 2019, https://www.eluniversal .com.mx/english/calderon-and-pena-nieto-spent-675-million-planes-and -helicopters.

NAACP. "Criminal Justice Fact Sheet." Accessed June 25, 2020, https://www .naacp.org/criminal-justice-fact-sheet/.

National Institute of Diabetes and Digestive and Kidney Diseases. "Overweight & Obesity Statistics." August 2017, https://www.niddk.nih.gov/health-information/health-statistics/overweight-obesity?dkrd=hispt0880.

Nature, "Local Spotlight. Rio de Janeiro, Brazil—Measuring Biodiversity and Ecological Integrity Benefits." Accessed June 22, 2020, https://www.nature.org/content/dam/tnc/nature/en/photos/LocalSpotlight_RioDeJaneiro_Brazil.pdf.

Nature Conservancy. "The Atlantic Forest." Accessed June 22, 2020, https://www.nature.org/en-us/get-involved/how-to-help/places-we-protect/atlantic-forest/.

Ndebele, Lenin. "Zim Soldiers Close to Starving." *Times Live*, November 10, 2019, https://www.timeslive.co.za/sunday-times/news/2019-11-10-zim-soldiers-close-to-starving/.

Neslen, Arthur. "Wind Power Generates 140% of Denmark's Electricity Demand." *The Guardian*, July 10, 2015, https://www.theguardian.com/environment/2015/jul/10/denmark-wind-windfarm-power-exceed-electricity-demand.

Nest, Michael. *Coltan*. Cambridge: Polity, 2011.

New Internationalist. "Hunger Myths." December 5, 1992, https://newint.org/features/1992/12/05/hunger.

Nunez, Christina. "Desertification, Explained." *National Geographic*, May 31, 2019, https://www.nationalgeographic.com/environment/habitats/desertification/.

Nyc.gov. "Climate Week: Solar Power in NYC Nearly Quadrupled since Mayor de Blasio Took Office and Administration Expands Target." *NYC*, September 23, 2016, https://www1.nyc.gov/office- of-the-mayor/news/767-16/climate-week-solar-power-nyc-nearly-quadrupled-since-mayor-de-blasio- took-office-and.

Olivero, J., J. E. Fa, and R. Real et al., "Recent Loss of Closed Forests Is Associated with Ebola Virus Disease Outbreaks," *Sci Rep* 7, no. 14291 (2017), https://doi.org/10.1038/s41598-017-14727-9.

Olympic.org. "The Refugee Olympic Team, A Symbol of Hope." April 5, 2017, https://www.olympic.org/news/the-refugee-olympic-team-a-symbol-of-hope.

Orangutan Foundation. "The Effects of Palm Oil." Accessed June 22, 2020, https://orangutan.org/rainforest/the-effects-of-palm-oil/.

Osborne, Hilary. "Virgin awarded almost £2bn of NHS contracts in the past five years." *The Guardian*, August 5, 2018, https://www.theguardian.com/society/2018/aug/05/virgin-awarded-almost-2bn-of-nhs-contracts-in-the-past-five-years.

Osborne, Samuel. "France's Summer Heatwave Killed 1,500 People, Health Minister Announces." *The Independent*, September 8, 2019, https://www

.independent.co.uk/news/world/europe/france-heatwave-summer-dead
-death-toll-europe-a9096806.html.

Oschmann, Roswitha. "Sea Watch Captain Speaks in Bad Honnef." *General Anzeiger*, April 5, 2019, https://www.general-anzeiger-bonn.de/region/sie
bengebirge/bad-honnef/sea-watch-kapitaenin-spricht-in-bad-honnef_aid
-44029639.

Oxfam. "Poverty in the USA." Accessed June 23, 2020, https://policy-practice
.oxfamamerica.org/work/poverty-in-the-us/.

Oxfam International. *An Economy for the 1%: How Privilege and Power in the Economy Drive Extreme Inequality and How This Can Be Stopped.* Oxford: Oxfam GB, https://s3.amazonaws.com/oxfam-us/www/static/media/files
/bp210-economy-one-percent-tax-havens-180116-en_0.pdf.

Palacios, Braulio. "Guatemala mejora posición en 'Top 10' de países exportado-res de café." *Republica*, December 5, 2019, https://republica.gt/2019/12/05
/guatemala-posicion-exportadores-cafe/.

Parkinson, Charles. "The Flow of Drugs and Blood in the Amazon Tri-Border Region." InSight Crime, April 4, 2014, https://www.insightcrime.org/news
/analysis/the-flow-of-drugs-and-blood-in-the-amazon-tri-border-region/.

Paterson, Tony. "Ursula von der Leyen: Is This the Next Woman to Become Chancellor of Germany?" *The Independent*, December 15, 2013, https://
www.independent.co.uk/news/world/europe/ursula-von-der-leyen-is-this
-the-next-woman-to-become-chancellor-of-germany-9006358.html.

Payton, Matt. "French Supermarkets Banned from Throwing Away and Spoiling Unsold Food." *The Independent*, February 5, 2016, https://www
.independent.co.uk/news/world/europe/french-law-bans-supermarkets
-throwing-away-and-spoiling-unsold-food-giving-them-to-food-banks
-and-a6855371.html.

Pengue, Walter. "The Impact of Soybean Expansion in Argentina." *Grain*, September 24, 2001, https://www.grain.org/es/article/entries/292-the
-impact-of-soybean-expansion-in-argentina.

Pensamiento y Acción Social, and Arbeitsgruppe Schweiz Kolumbien. *Shadow Report on the Sustainability of Glencore's Operations in Colombia.* Bogota, 2015.

Petrequin, Samuel. "EU Lays out 1 Trillion-Euro Plan to Support Green Deal." *AP News*, January 14, 2020, https://apnews.com/5d4db8ffda58f03f0
90a04c35f0a2dc8.

Piketty, Thomas. *Capital in the Twenty-First Century.* Translated by Arthur Goldhammer. Cambridge, MA; London, England: Belknap Press of Harvard University Press, 2017.

Piper, Kelsey. "The Climate Renegade." *Vox*, June 4, 2019, https://www.vox
.com/the-highlight/2019/5/24/18273198/climate-change-russ-george-uni
lateral-geoengineering.

Pitts-Singer, Theresa L., and Rosalind R. James. "Bees in Nature and on the Farm." In *Bee Pollination in Agricultural Ecosystems*, edited by Rosalind R. James and Theresa L. Pitts-Singer, 3–9. Oxford: Oxford University Press, 2008.

Polanco, Anggy. "Venezuela Probes Reports of Prison Cannibal Deaths." *Reuters*, October 17, 2016, https://www.reuters.com/article/us-venezuela-cannibalism-idUSKBN12H1SZ.

Ramboll. "Anholt Offshore Wind Farm—Denmark's Largest Offshore Wind Farm." Accessed June 25, 2020, https://ramboll.com/projects/re/anholt-offshore-wind-farm.

Rand, Ayn. *The Fountainhead*. New York: New American Library, 2016. Kindle.

Rappaport, Helen, and Marian Wright Edelman. *Encyclopedia of Women Social Reformers*, vol. 1. Santa Barbara: ABC-CLIO, 2001.

Regan, Helen. "Italy Announces Lockdown as Global Coronavirus Cases Surpass 105,000." *CNN*, March 8, 2020, https://edition.cnn.com/2020/03/08/asia/coronavirus-covid-19-update-intl-hnk/index.html.

Reporters Without Borders (RSF). "2020 World Press Freedom Index." Accessed June 20, 2020, https://rsf.org/en/ranking_table.

Reuters. "Colombia Coal Production up 6.5% in First Quarter." May 22, 2019, https://www.reuters.com/article/us-colombia-coal/colombia-coal-production-up-6-5-in-first-quarter-idUSKCN1SS2OA.

Reuters, Oslo. "Norway Wealth Fund Grows to Record 10 Trillion Crowns." *Reuters*, October 25, 2019, https://www.reuters.com/article/us-norway-swf-record/norway-wealth-fund-grows-to-record-10-trillion-crowns-idUSKBN1X41AO.

Reynolds, Jesse L. *The Governance of Solar Geoengineering: Managing Climate Change in the Anthropocene*. Cambridge: Cambridge University Press, 2019.

Rodionova, Zlata. "Denmark Reduces Food Waste by 25% in Five Years with the Help of One Woman—Selina Juul." *The Independent*. February 28, 2017, https://www.independent.co.uk/news/business/news/denmark-reduce-food-waste-25-per-cent-five-years-help-selina-juul-scandanavia-a7604061.html.

Rohrlich, Justin. "This Is How Much It Costs to Be Smuggled over the US Border." *Quartz*, June 10, 2019, https://qz.com/1632508/this-is-how-much-it-costs-to-cross-the-us-mexico-border-illegally/.

Rotman, David. "A Cheap and Easy Plan to Stop Global Warming." *MIT Technology Review*, February 8, 2013, https://www.technologyreview.com/2013/02/08/84239/a-cheap-and-easy-plan-to-stop-global-warming/.

Samuel, Sigal. "A Staggering 1 Billion Animals Are Now Estimated Dead in Australia's Fires." *Vox*, January 7, 2020, https://www.vox.com/future-perfect/2020/1/6/21051897/australia-fires-billion-animals-dead-estimate.

Santander. "Support for Santander Mexico Customers on Account of CO-VID-19." March 26, 2020, https://www.santander.com/content/dam/san
tander-com/en/documentos/actualidad/2020/03/no-2020-03-26-support
-for-santander-mexico-customers-on-account-of-covid-19-en.pdf.

Sarukhán, J., et al. *Capital natural de México. Síntesis: conocimiento actual, evaluación y perspectivas de sustentabilidad.* Mexico: Comisión Nacional para el Conocimiento y Uso de la Biodiversidad, 2009, http://bioteca.biodiversidad.gob.mx/janium/Documentos/14039.pdf.

Schücking, Heffa. *Banking on Coal.* Urgewald, BankTrack, CEE Bankwatch Network and Polska Zielona Sieć, 2013, https://www.banktrack.org/down
load/banking_on_coal/banking_on_coal_updated.pdf.

Schwägerl, Christian, "In Conservative Bavaria, Citizens Force Bold Action on Protecting Nature," *Yale Environment 360*, April 25, 2019, https://e360
.yale.edu/features/in-conservative-bavaria-citizens-force-bold-action-on
-protecting-nature.

Schwedhelm, Alejandro, Wei Li, Lucas Harms, and Claudia Adriazola-Steil. "Biking Provides a Critical Lifeline During the Coronavirus Crisis." *World Resources Institute*, April 17, 2020, https://www.wri.org/blog/2020/04
/coronavirus-biking-critical-in-cities.

Shelley, Fred M., and Reagan Metz. *Geography of Trafficking: From Drug Smuggling to Modern-Day Slavery.* Santa Barbara: ABC-CLIO, 2017.

Shiva, Vandana. "Ecological Reflections on the Coronavirus." Medium, March 23, 2020, https://medium.com/post-growth-institute/ecological
-reflections-on-the-coronavirus-93d50bbfe9db.

Shukman, David. "Electric Car Future May Depend on Deep Sea Mining." *BBC*, November 13, 2019, https://www.bbc.com/news/science-environ
ment-49759626.

Shumaker, Lisa. "U.S. coronavirus deaths exceed 70,000 as forecasting models predict grim summer." *Reuters*, May 5, 2020, https://www.reuters.com/ar-ticle/us-health-coronavirus-usa-casualties-idUSKBN22H2ES.

Simone-Finstrom, Michael, and Marla Spivak, "Propolis and Bee Health: The Natural History and Significance of Resin Use by Honey Bees," *Apidologie* 41, no. 3 (2010): 295–311, https://doi.org/10.1051/apido/2010016.

Simons, Eric. "California's Early June Heat Wave Cooked Coastal Mussels in Place." *Bay Nature*, June 26, 2019, https://baynature.org/2019/06/26
/californias-early-june-heat-wave-cooked-coastal-mussels-in-place/.

Simpson, Andrew, ed. *Language and National Identity in Africa.* Oxford: Oxford University Press, 2008.

Simpson, John. "Peru's 'Copper Mountain' in Chinese Hands." *BBC News*, June 17, 2018, http://news.bbc.co.uk/2/hi/americas/7460364.stm.

Steinmetz, Hillel. "Noam Chomsky Warns of Sixth Extinction." *Chicago Maroon*, September 30, 2016, https://www.chicagomaroon.com/article/2016/9/30/noam-chomsky-warns-sixth-extinction/.

Stock, Matthew. "Air Pollution Plunges in European Cities amid Coronavirus Lockdown: Satellite Data." *Reuters*, March 27, 2020, https://www.reuters.com/article/us-health-coronavirus-europe-pollution/air-pollution-plunges-in-european-cities-amid-coronavirus-lockdown-satellite-data-idUSKBN21E2UK.

Stop Wasting Food. "Stop Wasting Food." Accessed June 23, 2020, https://stopwastingfoodmovement.org/.

Strochlic, Nina. "Six Years Ago, Boko Haram Kidnapped 276 Schoolgirls. Where Are They Now?" *National Geographic*, accessed June 26, 2020, https://www.nationalgeographic.com/magazine/2020/03/six-years-ago-boko-haram-kidnapped-276-schoolgirls-where-are-they-now/.

Strupczewski, Jan, and Gabriela Baczynska. "EU Leaves Poland out of 2050 Climate Deal after Standoff." *Reuters*, December 11, 2019, https://www.reuters.com/article/us-climate-change-eu/eu-leaves-poland-out-of-2050-climate-deal-after-standoff-idUSKBN1YG01I.

Struzik, Ed. "Shipping Plans Grow as Arctic Ice Fades." *Yale Environment 360*, November 17, 2016, https://e360.yale.edu/features/cargo_shipping_in_the_arctic_declining_sea_ice.

Süddeutsche Zeitung. "We Don't Regret Anything." Accessed June 20, 2020, https://panamapapers.sueddeutsche.de/articles/57162ccca1bb8d3c3495bc5e/.

Teather, David. "Enron Accountant Pleads Guilty." *The Guardian*, December 29, 2005, https://www.theguardian.com/business/2005/dec/29/corporatefraud.enron.

The Telegraph. "14 of the World's Most Beautiful Fountains (Drought Permitting)—Dubai Fountain." July 26, 2017, https://www.telegraph.co.uk/travel/galleries/most-beautiful-fountains/dubai-fountain/.

Thobhani, Akbarali. *Western Sahara since 1975 under Moroccan Administration.* Lewiston, NY: Edwin Mellen Press, 2002.

Thomas, Phil. "Coronavirus: US Surpasses Italy for Most Recorded Deaths in the World with 20,254." *The Independent*, April 11, 2020, https://www.independent.co.uk/news/world/americas/coronavirus-us-most-confirmed-deaths-latest-covid-19-a9460866.html.

Tidman, Zoe. "A British Citizen Emits More CO2 in Two Weeks Than Some People in Africa Do in a Year, Research Shows." *The Independent*, January 6, 2020, https://www.independent.co.uk/environment/british-carbon-footprint-africa-emissions-oxfam-climate-change-a9271861.html.

Time Staff. "Here's Donald Trump's Presidential Announcement Speech." *Time*, June 16, 2015, https://time.com/3923128/donald-trump-announce ment-speech/.

Transparency International. *Corruption Perceptions Index 2019*. Transparency International, 2019. Accessed June 20, 2020, https://images.transparency cdn.org/images/2019_CPI_Report_EN_200331_141425.pdf.

Tremlett, Giles. "Spain's King Juan Carlos under Fire over Elephant Hunt- ing Trip." *The Guardian*, April 15, 2012, https://www.theguardian.com /world/2012/apr/15/spain-king-juan-carlos-hunting.

Trussell Trust. "About." Accessed June 22, 2020, https://www.trusselltrust .org/about/.

———. "Food Bank Statistics for Previous Financial Years with Regional Breakdown 2018–2019." Accessed June 20, 2020, https://www.trusselltrust .org/news-and-blog/latest-stats/end-year-stats/#fy-2018-2019.

———. "Food Bank Statistics for Previous Financial Years with Regional Breakdown 2017–2018." Accessed June 22, 2020, https://www.trusselltrust .org/news-and-blog/latest-stats/end-year-stats/#fy-2017-2018.

Turner, Simon. *Beekeeping as a Business: A Practical Guide to Beekeeping in Uganda*, 4th ed. Kampala, Uganda: Malaika Honey, 2018.

UNHCR. "79.5 Million Forcibly Displaced People Worldwide at the End of 2019." June 18, 2020, https://www.unhcr.org/figures-at-a-glance.html.

United Nations. *Kyoto Protocol to the United Nations Framework Convention on Climate Change*. New York: United Nations, 1998, https://unfccc.int /resource/docs/convkp/kpeng.pdf.

United Nations, Convention on Climate Change. *The Kyoto Protocol—Status of Ratification*. Accessed June 20, 2020, https://unfccc.int/process/the-kyoto -protocol/status-of-ratification.

United Nations, Department of Economic and Social Affairs. *World Economic and Social Survey 2011: The Great Green Technological Transformation*. New York: United Nations, 2011, https://www.un.org/development/desa/dpad /wp-content/uploads/sites/45/2011wess.pdf.

United Nations, Department of Economic and Social Affairs, Population Divi- sion. *World Population Prospects: The 2012 Revision, Volume I: Comprehensive Tables ST/ESA/SER.A/336*. 2013, https://population.un.org/wpp/Publica tions/Files/WPP2012_Volume-I_Comprehensive-Tables.pdf.

———. *World Population Prospects: The 2017 Revision, Key Findings and Ad- vance Tables. Working Paper No. ESA/P/WP/248*. 2017, https://population .un.org/wpp/Publications/Files/WPP2017_KeyFindings.pdf.

———. *World Population Prospects 2019*. New York: United Nations, 2019. Online ed., rev. 1, https://population.un.org/wpp/Download/Files/1_Indi cators%20(Standard)/EXCEL_FILES/1_Population/WPP2019_POP _F01_1_TOTAL_POPULATION_BOTH_SEXES.xlsx.

———. *The World's Cities in 2018—Data Booklet.* New York: United Nations, 2018, 4 and 26, https://www.un.org/en/events/citiesday/assets/pdf/the_worlds_cities_in_2018_data_booklet.pdf.

United Nations Office on Drugs and Crime (UNODC). "Heroin and Cocaine Prices in Europe and USA." *United Nations.* Accessed June 24, 2020, https://dataunodc.un.org/drugs/heroin_and_cocaine_prices_in_eu_and_usa-2017.

United Nations Security Council Meeting. "Humanitarian Crisis in Democratic Republic of Congo Will Worsen Without Political Transition, End to Violence, Speakers Warn Security Council." March 19, 2018, https://www.un.org/press/en/2018/sc13253.doc.htm.

Vale. *A Year of Extraordinary Performance: Annual Report 2010.* Vale S. A., 2011, http://www.vale.com/EN/investors/information-market/annual-reports/20f/20FDocs/20F_2010_i.pdf.

van Wyk, Chris. *Helen Joseph.* South Africa: Awareness Publishing, 2003.

Vernon Trail, Jesse. "The Truth about Peat Moss." *The Ecologist*, January 25, 2013, https://theecologist.org/2013/jan/25/truth-about-peat-moss.

Wacket, Marcus. "Germany to Phase out Coal by 2038 in Move away from Fossil Fuels." *Reuters*, January 25, 2019, https://www.reuters.com/article/us-germany-energy-coal/germany-to-phase-out-coal-by-2038-in-move-away-from-fossil-fuels-idUSKCN1PK04L.

Welt. "Beekeepers Are Concerned about Pesticide Residues in Bee Pollen." November 18, 2019, https://www.welt.de/regionales/rheinland-pfalz-saarland/article203599438/Imker-besorgt-ueber-Rueckstaende-von-Pestiziden-in-Bienenpollen.html?fbclid=IwAR3c_q0dKY7L8wjY444qNT0wPZjBT5asRknDZ6r5noX9Ow9t5f7-4H80jQ0.

Whiteman, Hilary. "India Heat Wave Kills 2,330 People as Millions Wait for Rain." *CNN*, June 2, 2015, https://edition.cnn.com/2015/06/01/asia/india-heat-wave-deaths/index.html.

Winkless, Laurie. "Dutch Trains Are Now Powered by Wind." *Forbes*, January 12, 2017, https://www.forbes.com/sites/lauriewinkless/2017/01/12/dutch-trains-are-now-powered-by-wind/#631148752d29.

Winston, Mark L. *Biology of the Honey Bee.* Cambridge, MA; London, England: Harvard University Press, 1991.

Wintour, Patrick. "Emmanuel Macron: 'More Choice Would Mean Fewer Children in Africa.'" *The Guardian*, September 26, 2018, https://www.theguardian.com/global-development/2018/sep/26/education-family-planning-key-africa-future-emmanuel-macron-un-general-assembly.

Wollheim Memorial. "Zyklon B: An Insecticide Becomes a Means for Mass Murder." Accessed June 26, 2020, http://www.wollheim-memorial.de/en/zyklon_b_en_2.

Wood, Zoe. "Fast Food Chains Prosper as Cash-Strapped Consumers Shun Retailers." *The Guardian*, January 29, 2012, https://www.theguardian.com /business/2012/jan/29/fast-food-chains-prosper-retailers-under-pressure.

———. "The World's 50 Best Restaurants 2017." 2017, https://www .theworlds50best.com/previous-list/2017.

World Health Organization. "Brazil." Accessed June 25, 2020, https://www .who.int/diabetes/country-profiles/bra_en.pdf.

———. "Ebola in the Democratic Republic of the Congo." Accessed June 26, 2020, https://www.who.int/emergencies/diseases/ebola/drc-2019.

World Meteorological Organization. "WMO Confirms 2019 as Second Hottest Year on Record." January 15, 2020, https://public.wmo.int/en/media /press-release/wmo-confirms-2019-second-hottest-year-record.

World Nuclear Association. "Nuclear Power in France." Accessed June 25, 2020, https://www.world-nuclear.org/information-library/country-profiles /countries-a-f/france.aspx.

World's 50 Best Restaurants. "Sustainable Restaurant Award." 2019, https:// www.theworlds50best.com/awards/sustainable-restaurant-award.

World Wind Energy Association (WWEA). "Denmark." *WWEA Policy Paper Series*, April 2018, http://www.wwindea.org/wp-content/uploads/2018/06 Denmark_full.pdf.

Wu, Xiao, Rachel C. Nethery, Benjamin M. Sabath, Danielle Braun, and Francesca Dominici. "Exposure to Air Pollution and COVID-19 Mortality in the United States." Department of Biostatistics, Harvard T. H. Chan School of Public Health, April 5, 2020, https://projects.iq.harvard.edu/files /covid-pm/files/pm_and_covid_mortality.pdf.

WWF. "8 Things to Know about Palm Oil." January 17, 2020, https://www .wwf.org.uk/updates/8-things-know-about-palm-oil.

———. "Amazon Deforestation." Accessed June 22, 2020, https://wwf.panda .org/our_work/forests/deforestation_fronts2/deforestation_in_the_amazon/.

———. "Second Only to the Amazon." Accessed June 22, 2020, https://wwf .panda.org/knowledge_hub/where_we_work/atlantic_forests/.

Index

Europe, 2, 5, 19, 21, 31, 53, 57, 66, 68, 71, 74, 83, 84, 93, 99, 103–105, 108–10, 112–20, 125, 128, 129, 134, 140, 143, 145, 150, 162, 175, 184, 186, 187, 189–91, 198, 204, 210, 215, 220, 223, 227, 228
European Commission, 223
European honeybee (*Apis mellifera*), 126–30
European hornet (*Vespa crabro*), 128
European Union, 92, 111, 114, 117, 131, 162
exploration, 28, 183, 213
exports, 21–24, 35, 43, 61, 74, 114, 116, 148, 149, 184, 187, 222
extinction, 2, 18, 47, 128, 135. *See also* sixth mass extinction
extraction, 8, 28, 75, 76, 84, 118, 182; coal, 36, 117; cobalt, 75, 184; lithium, 74; mineral, 185; oil, 4, 35, 220. *See also* miners
Exxon, 81

fair trade, 8, 22, 24, 25, 41, 75, 88, 101, 134, 184
false codling moth (*Thaumatotibia (Cryptophlebia) leucotreta*), 149
famine, 1, 184, 185, 188
Farage, Nigel, 113, 114, 117
farm, 8, 46, 59, 60, 88, 94, 131, 139, 141–43, 149, 153, 157, 158, 160, 161, 209, 229, 235; ecological, 154–56, 181; rooftop, 90, 122; urban, 87, 90
farmer, 8, 20–23, 25, 27, 37, 41, 57, 60, 61, 62, 131, 142, 143, 145, 146, 148, 149, 153–57, 159, 160, 162, 163, 185, 192, 202, 231
farming, 8, 16, 43–45, 159, 192; cattle, 7, 44, 45; coffee, 46;

ecological, 8, 154–56; industrial, 61, 62, 156, 192; livestock, 17, 19, 38; organic, 24, 159; peasant, 57, 61, 89, 160; school, 155–57. *See also* cows
fat, 28, 89
Favela da Paz, 49–52
Feedback (food activism organization), 90
Feeding the 5,000 (story), 90, 92
femicide, 21
fertilizers, 16, 50, 155; organic, 24
field marigold (*Calendula arvensis*), 126
financial markets, 23, 31, 207, 291
Fipronil, 133
fish, 7, 17, 37, 43, 90, 84, 165
flood, 2, 8, 154, 162, 164, 185
Florida, 74, 77, 80
Floyd, George, 192
food, 1, 8, 25, 28, 38, 41, 42, 43, 51, 59, 62, 81, 85, 86–94, 109, 112, 120, 122, 126–31, 133, 134, 142, 145, 147, 160, 163–66, 169, 170, 173, 178, 184, 191, 192, 197, 198, 208, 209, 229, 234; banks, 92; fast, 89; GMO, 131, 159; hyper-local, 86, 87, 93; organic, 75, 88, 90, 93, 131, 134; sustainable, 93–95; waste, 49, 87, 90, 91–95, 164
foreign exchange, 2, 23, 29, 167
fracking, 80
France, 93, 105, 111, 113, 117, 154, 163, 197, 206, 223
Fridays for Future, 223
Froidevaux, Sandra, 36
Fuentes, Paty, 122
Fukushima, 212
fumigation, 157

About the Author

Osseily Hanna was born in London in 1978. He is a writer, photographer, and filmmaker with experience in more than forty countries. He combines his science and banking background with empathy, dedication, and passion for humanity and nature to write books and direct documentary films about the most pressing issues of our times. He gave up a successful career in global financial markets in 2011 to write his first book, *Music and Coexistence*, which was first published in English by Rowman & Littlefield.

CPSIA information can be obtained
at www.ICGtesting.com
Printed in the USA
BVHW080414310321
603749BV00001B/1